God's Hidden Creation Numbers*

Functions of the universe!

Charles D Landis

AuthorHouse™
1663 Liberty Drive, Suite 200
Bloomington, IN 47403
www.authorhouse.com
Phone: 1-800-839-8640

©*2008 Charles D Landis. All rights reserved.*

No part of this book may be reproduced, stored in a retrieval system, or transmitted by any means without the written permission of the author.

First published by AuthorHouse   7/02/2008

ISBN: 978-1-4343-9692-1 (sc)
ISBN: 978-1-4343-9693-8 (hc)
ISBN: 978-1-4343-9691-4 (e)

Printed in the United States of America
Bloomington, Indiana

This book is printed on acid-free paper.

"In the beginning God created the heaven and the earth" Genesis 1:1

"In the beginning was the Word, and the Word was with God, and the Word was God" John 1:1

Introduction

Is there one common number that seems to be the essence of the universe? What numbers could possibly bring together these physical and mathematical properties and Biblical concepts?

- Relationships between the physical and mathematical constants
- The Fine Structure Constant
- DNA
- The atomic weights of the 92 natural elements
- The Speed of Light in a Vacuum
- Music vibration rates
- The Prime numbers
- Pi and Euler's e and Phi (ϕ) (The Golden Ratio)
- The Earth's physical constants
- The Hebrew and Greek and English alphabets
- The Holy Scriptures of the Bible
- And much more……..

This book is a treasure hunt for this elusive number. It is not possible to cover the entire universe of physical, chemical, biological, mathematical, astronomical and Biblical topics in this small book. Only a small portion of the topics can be covered. But I am convinced that these elusive numbers, i.e. functions of the universe, have been found. Hopefully, this book will stimulate scientists and mathematicians and Biblical scholars to do further study and research to validate these claims.

The mathematics used in this book is simple eighth grade level arithmetic; it is not a book of theories, but actual mathematical calculations that are easy to understand.

I am convinced that complex theories are not needed to truly understand the construction of the universe. While my mathematical calculations are simple, yet the principles of cryptology may be needed sometimes to uncover these elusive numbers. There does not seem to be just one simple type of calculation needed to uncover these numbers, but sometimes you need to "play around with" various types of calculations in order to get the answer.

I am convinced that a knowledge and commitment to the "Holy Scriptures of the Bible" are absolutely essential in understanding these "hidden numbers of the universe". I do not think it is possible to be committed to an "evolutionary world view" when you see the precise mathematical construction of our universe. This book makes a strong case, I think, for a "creationist world view". The universe could not have possibly come together with a "big bang" or an "evolutionary process". Each person has their own opinion, of course. Some readers will not like the results shown in this book simply because it does not square with their perceived understanding; but the numerical calculations in this book should settle the matter. I only ask that the readers of this book keep an open mind to the mathematical evidence given in this book. It is reasonable to assume that a Creator God would "leave His handwriting all over the creation" even as "retail merchandise usually tags the nation of manufacture". In this book we will see numerous examples of "His handwriting" which also gives His identity and Name.

Let's get started on this exciting journey!!

The Hebrew and Greek and English alphabets

The Hebrews and the Greeks used their alphabets in their written languages and in their numeral systems. A good book that explains this and features the Hebrew and Greek alphabets is James Harrison's book, "The Pattern & the Prophecy: God's Great Code". Throughout our study here, we will be "spelling out the letter names" and using the "numeric values" of these letters of the alphabet. Please note that this book is not a study in gematria, but in simply using known Hebrew, Greek and English letter values and showing their relationships with physical and mathematical constants and values. Listed below are several charts illustrating this information:

Common Numeric Values for the Hebrew, Greek and English Alphabets

Hebrew Alphabet:

Aleph (1)	Yod (10)	Koph (100)
Beth (2)	Kaph (20)	Resh (200)
Gimmel (3)	Lamed (30)	Shin (300)
Daleth (4)	Mem (40)	Tau (400)
He (5)	Num (50)	Koph (500)
Vau (6)	Samech (60)	Mem (600)
Zayin (7)	Ayin (70)	Num (700)
Cheth (8)	Pe (80)	Pe (800)
Teth (9)	Tsaddi (90)	Tsaddi (900)

The Sum of the letter numeric values = 4,995 (dropping the zeros) = 27 x 185 = 37 x 135. And 4995 − 4599 = 396, where 8 x 396 = 3168 = "Lord Jesus Christ" in Greek numeric value.

Also, $(4,995)^2$ = 24,950,025. When we multiply these digits together, we get: 3,600. Then $(3,600)^2$ = 12,960,000, where 250 x 51,840 = 12,960,000 and 1,250 x 10,368 = 12,960,000. Both of these numbers (51,840 and 10,368) play a very important role in "bringing together the components of the universe" as we shall see throughout this book.

4995 − 3168 = 1827, where 3 x 1827 = 5481, which is one of twenty-four permutations of 5184, which is related to the "Atomic Weights of all 92 Natural Elements". There are twenty-four permutations of 4995 = 179,982 = 18 x 9999 = 27 x 6666.

Here are the twenty-four (24) permutations of 5184 for your reference through out this book:

5184	1548	4158	8145
5148	1584	4185	8154
5418	1458	4518	8451
5481	1485	4581	8415
5814	1845	4815	8514
5841	1854	4851	8541

Greek Alphabet:

Alpha (1)	Iota (10)	Rho (100)
Beta (2)	Kappa (20)	Sigma (200)
Gamma (3)	Lambda (30)	Tau (300)
Delta (4)	Mu (40)	Upsilon (400)
Epsilon (5)	Nu (50)	Phi (500)
Stigma (6)	Xi (60)	Chi (600)
Zeta (7)	Omicron (70)	Psi (700)
Eta (8)	Pi (80)	Omega (800)
Theta (9)	Koppa (90)	Sampsi (900)

The Sum of the letter numeric values = 4,995. The same calculations apply as for the Hebrew numeric values above.

English Alphabet:

A (6)	B (12)	C (18)
D (24)	E (30)	F (36)
H (48)	I (54)	J (60)
K (66)	L (72)	M (78)
N (84)	O (90)	P (96)
Q (102)	R (108)	S (114)
T (120)	U (126)	V (132)
W (138)	X (144)	Y (150)
Z (156)		

Please note that a number of researchers have validated this numbering scheme for the English alphabet and my additional calculations will show the relationships between these three (3) alphabet systems.

The Sum of the letter numeric values = 2106, where $(2,106)^2$ = 4,435,236 = 0.5184 x 8,555,625 = 0.10368 x 42,778,125.

2106 + 1062 = 3168 and 3 x 2106 = 6318, where 6318 and 1062 are permutations of 3168 and 2106 respectively.

The twenty-four permutations of 2106 = 59,994 = 6 x 9999 = 27 x 2222; thus we can see that the English numeric letter values are one-third the values for the Hebrew and Greek alphabets. And $\frac{179,982}{59,994} = 3$. Therefore, we should be able to substitute any of these alpha numeric systems with similar calculations.

These two numbers (51840 and 10,368) and their many permutations continue to show up in all aspects of our created universe. We see from the above illustrations that the Hebrew and Greek and English alphabets are directly related to each other. Therefore, I will be using the English letters and their numeric values throughout this book so that those persons who are not knowledgeable in Hebrew and Greek can understand this book.

To further illustrate this connection between these three alphabets, please review the following chart, where the "spelling of the Hebrew letters" and the "spelling of the Greek letters" are used to calculate the English numeric values for each of these letters:

Hebrew Alphabet and Greek Alphabet with English Numeric Calculated Values

Hebrew	Greek
Aleph = 114	Alpha = 110
Beth = 215	Beta = 208
Gimel = 91	Gamma = 89
Daleth = 248	Delta = 240
He = 13	Epsilon = 324
Vau = 701	Stigma = 357
Zayin = 1560	Zeta = 1006
Cheth = 224	Eta = 206
Teth = 413	Theta = 414
Yod = 764	Iota = 270
Kaph = 99	Kappa = 162
Lamed = 80	Lambda = 78
Mem = 85	Mu = 340
Num = 390	Nu = 350
Samech = 157	Xi = 609
Ayin = 760	Omicron = 312
Pe = 75	Pi = 79
Tsaddi = 318	Koppa = 221
Koph = 158	Rho = 158
Resh = 203	Sigma = 157
Shin = 167	Tau = 501
Tau = 501	Upsilon = 619

Hebrew Alphabet and Greek Alphabet with English Numeric Calculated Values

(Continued)

Hebrew	Greek
Koph = 158	Phi = 87
Mem = 85	Chi = 20
Num = 390	Psi = 179
Pe = 75	Omega = 113
Tsaddi = 318	Sampsi = 320
Total numeric Value = 8362	Total Numeric Value = 7529

Note that $(8362)^2 = 69{,}923{,}044$ where the product of these digits = 46,656 and 9 x 5184 = 46,656 and 45 x 10368 = 466,560.

And $(7529)^2 = 56{,}685{,}841$, where the product of these digits = 345,600. Also, $(3456)^2 = 11{,}943{,}936$, where 2304 x 5184 = 11,943,936 and 1152 x 10368 = 11,943,936.

Thus, we see again how the Hebrew and Greek and English alphabet numeric values coordinate to produce the values of 5184 and 10368.

It is important to note that permutations are simply a "rotation of the same digits" in any given number. For example, 3,168 has 24 permutations and 10,368 has 120 permutations and 51,840 has 120 permutations. I like to explain these permutations like this: Suppose you want to take a "family picture" during a holiday celebration, where there are 8 relatives to be included in the picture. You could take a total of 8! = 40,320 separate individual pictures, where each of these 8 relatives are "rotated into a different location" for the each of these 40,320 pictures.

And you could then call each of these pictures a "family portrait" because the same persons are in each photograph, but standing in a different location. The "essence" of each photo is the same; i.e. a "family portrait", but each photograph is a distinctly different picture. Thus, 10,368 has the same "essence" as 03,168 because the actual digits have simply been re-positioned. Likewise, 51,840 has the same "essence" as 1,548. This "re-positioning of the digits" is frequently found throughout our created universe.

It seems that the Creator uses certain techniques to "hide" His creation numbers; He seems to use "cryptology" to "hide" His numbers from persons who do not believe in Him and who do not diligently seek Him.

Also, throughout this book we will eliminate "decimal points" and "zeros" many times because we are really looking to evaluate the actual digits 1, 2, 3, 4, 5, 6, 7, 8, and 9 as they appear in various permutations within the physical and mathematical constants. For example, when we evaluate Pi = 3.1415926535, we simply consider the value of 31415926535 in our hunt for this creation number. Eliminating these decimal points and zeros does not degrade or invalidate our calculations.

In this section we will look at the "Holy Scriptures of the Bible" to see how Genesis 1:1 "In the beginning God created the heaven and the earth" relate directly to these "hidden numbers" of creation.

Genesis 1:1 "In the beginning God created the heaven and the earth"

In this section we will investigate Genesis 1:1 "In the beginning God created the heaven and the earth" (KJV) to show that the numbers 5184 and 10368 (and their various permutations) are clearly found in this creation scripture.

In addition, we will show the "title" and "name" of God and His relationship with the "Speed of Light" and Pi. In later chapters we will show many more relationships. Please notice all of the numerics which relate directly to the Creator God.

Listed here are 21 digits of Pi = 3.14159265358979323846. In Genesis 1:1 we have seven (7) Hebrew words with the following numeric values: "In the beginning" = 913 and "created" = 203 and "God" = 86 and "aleph tav" = 401 and "heaven" = 395 and "and" = 407 and "earth" = 296. Thus, we take these numeric values, in the order in which they appear in Hebrew, and we get 91320386401395407296 (20 digits) without any spaces between them.

Now we can find the Pi digits in Genesis 1:1:

    3.1415926_3_897932__46 Thus, we get a 7 digit accuracy using Genesis 1:1 digits to find Pi; all the digits of Genesis 1:1 were used (each digit was only used once) except 3 zeros (000). If we eliminate the "gaps" in the above Pi value by "removing the gaps", then:

    3.1415926535897932 Minus 3.1415926389793246, we get a difference of just 0.0000000146105, which shows that Genesis 1:1 digits provide an extremely accurate value of Pi.

We can also use the Trinity Function in evaluating Genesis 1:1 numeric values. Let me illustrate how the Trinity Function is calculated.

Taking the digits of Genesis 1:1, we get: $9^3 + 1^3 + 3^3 + 2^3 + 3^3 + 8^3 + 6^3 + 4^3 + 1^3 + 3^3 + 9^3 + 5^3 + 4^3 + 7^3 + 2^3 + 9^3 + 6^3 =$ 3826, which happens to be one of 24 permutations of 2368 = "Jesus Christ" in Greek numerics! Please note: You need to apply the Trinity Function repeatedly on the end result (in case 3826) until you get a terminal non-repeating number.

In this particular case, the first application of the Trinity Function gave us the "name" of "God" the "Creator" in just one application. However, when we repeat the Trinity Function on the number 3826, we get: $3^3 + 8^3 + 2^3 + 6^3 = 763$. Again, TF (763) = $7^3 + 6^3 + 3^3 = 586$.

Again, TF (586) = $5^3 + 8^3 + 6^3 = 853$. Again, TF (853) = $8^3 + 5^3 + 3^3 = 664$. Again, TF (664) = $6^3 + 6^3 + 4^3 = 496$. Again, TF (496) = $4^3 + 9^3 + 6^3 = 1009$. Again, TF (1009) = $1^3 + 9^3 = 730$. Again, TF (730) = 370 a terminal, non-repeating number. Please refer this illustration in later pages if you want to remember how the Trinity Function operates on any given number.

Jesus Christ, the Aleph Tav

Genesis 1:1 "In the beginning God created the heaven and the earth".

Hebrew Numeric Values

913 = in the beginning

203 = created

86 = God

401 = aleph tav   Aleph Tav = Jesus Christ

395 = heavens

407 = and

296 = earth

Total = 2701= 37 x 73   37 and 73 are both Prime Numbers.

2701 is 37th hexagon number   and 2701 is the 73rd triangle number

Genesis 1:1 "In the beginning God created the heaven and the earth".

Reverse Numeric Values

319 = in the beginning

302 = created

68 = God

104 = aleph tav

593 = heavens

704 = and

296 = earth (Not reversed because this is man's domain)

Total = 2386   2368 = 888 (Jesus in Greek numerics) + 1480 (Christ in Greek numerics). Note: God (Hebrew) = 86 and God (Reverse Hebrew) = 68

Then 86 - 68 = 18 and 2386 – 18 = 2368 (Jesus Christ in Greek)

Note: Hebrew words and numbers read from right to left, whereas the Greek and other languages read from left to right. Thus, "Aleph Tav" is the Name of Jesus Christ in Hebrew and Jesus Christ is found with Greek numeric value in Reverse Hebrew with His Name "Jesus Christ" spelled in English! Thus, God seems to co-relate the Hebrew, Greek and English languages.

Interesting Relationships Found in Genesis 1:1

2368 = 37 x 64 = Power x Truth in Greek numerics

2701 – 2386 = 315, which is a permutation of 153, "the image of Jesus" in Greek numeric using the Trinity Formula

2701 – 2368 = 333 = 9 X 37, which shows "Divine Completeness" X "Power" in Biblical numbers

Hebrew numeric shows 86 (God) + 401 (Aleph Tav) = 487

Reverse Hebrew numeric shows 68 (God) + 104 (Aleph Tav) = 172 and 172 – 86 = 86

487 – 172 = 315 and 401 -86 = 315, where 315 is a permutation of 153, the "image of Jesus" in Greek numeric using the Trinity Formula.

104 – 68 = 36 = $1^3 + 2^3 + 3^3$, where 1 represents God, the Father and 2 represents the Son and 3 represents the Holy Spirit. Note: 36 is a "Triangular Number" and a "Square Number".

Also, $6 = \sqrt[2]{(1^3 + 2^3 + 3^3)}$; therefore

$86 - 68 = 18 = (6 + 6 + 6) = [\sqrt[2]{(1^3 + 2^3 + 3^3)} + \sqrt[2]{(1^3 + 2^3 + 3^3)} + \sqrt[2]{(1^3 + 2^3 + 3^3)}]$

Note: 18 is equal to the sum of its digits of its cube: $18^3 = 5832$

Note: 104 is semi-perfect, because it is the sum of some of its own divisors: $104 = 52 + 26 + 13 + 8 + 4 + 1$. Note: $407 = 4^3 + 0^3 + 7^3$ Note: Number 73: All integers can be represented as the sum of at most 73 6th powers.

Note: Number 37: Every number is the sum of at most 37 5th powers. Any 3-digit multiple of 37 remains a multiple when its digits are cyclically permuted.

"In the beginning God created the heaven and the earth" Genesis 1:1

The Hebrew numeric values for this verse are as follows:

913 203 86 401 395 407 296. If we "sum the cubes of each digit in this series" we get the Sum Total = 3826, which is just one of twenty-four permutations of 2368 = "Jesus Christ" in Greek numerics.

When we apply the Trinity Function to the "Genesis 1:1 digits", we get 3826, which is just one of 24 permutations of 2368, which is "Jesus Christ" in Greek numerics. The final iteration of the Trinity Function on the "digits of Genesis 1:1" is 370.

When we apply the Trinity Function to the 21 digits of Pi (see above), we get 250, 133, 55 as endlessly repeating iterations of the Trinity Function; in other words, Pi is a "repeating, cyclical and never ending series of digits", which confirm many other mathematicians research.

"Speed of Light in a vacuum" = 299,792,458 $\frac{meters}{second^{-1}}$.

Find the "Speed of Light in a Vacuum" in Genesis 1:1 digits (See above). When we do this, there are exactly 11 digits left over that were not used: 1, 3, 0 3, 6, 0, 1, 3, 4, 0, 6.

If we take the Trinity Function of 13036013406 (as the pre-image), then we get 153 as the "image", which the "image of Jesus". Note: when we take the Greek numerics value of "Jesus" = 888 and apply the Trinity Function, we get 153).

Also, if we multiply the individual digits (left over above), we get: 1 x 3 x 3 x 6 x 1 x 3 x 4 x 6 = 3888 = 27 x 144. When we add up the value of all 24 permutations of 3888 we get a value of 174,312 and if we add up all 24 permutations of 3888 by reversing the digits in each permutation, we get a value of 329,247. Now subtract the value 174,312 from 329,247 to get a value of 503,559. Then take 5 x 3 x 5 x 5 x 9 = 3375.

3375 – 3168 = 207, where 207 = "Light" in Hebrew numerics and 3168 = "Lord Jesus Christ" in Greek numerics. Thus, we see that the "Lord Jesus Christ" is the "Spiritual Light of the World" and the "Creator of Physical Light" in this world.

The Greek numerics value of "Jesus Christ" = 2368. When we add up all 24 permutations of 2368 we get a value of 126,654. Then add up all 24 permutations of 2368 by reversing the digits in each permutation, we get a value of 166,551. Now 166,551 − 126,654 = 39,897. The 3 x 9 x 8 x 9 x 7 = 13608, where 03168 is just one permutation of 13608. Of course, 3168 = "Lord Jesus Christ" in Greek numerics.

If we multiply each digit of all 24 permutations of 2368, we get the value of 144,506,880,000.

The Trinity Function of 126,654 goes to 153 and the Trinity Function of 166,551 goes to 153 and the Trinity Function of 144,506,880,000 goes to 153, where 153 is the "image of Jesus" when the Greek numerics value of "Jesus" = 888 is applied to the Trinity Function.

Multiplying the digits of 913 203 86 401 395 407 296 as if it were one large number, we get: 12,697,896,960 = 4 x 10,368 x 306,180, where $\frac{10,368}{2}$ = 5184 and 03168 is one of 120 permutations of 10368 and 003168 is just one of 720 permutations of 306,180.

If we divide the digits of 913 203 86 401 395 407 296 by their reverse order, we get:

$\frac{913203864013954 07296}{692704593104683 02319}$ = 1.3183164556785 and then we multiply these digits, we get: 1 x 3 x 1 x 8 x 3 x 1 x 6 x 4 x 5 x 5 x 6 x 7 x 8 x 5 = 72,576,000 = 7,000 x 10368 = 1,400 x 51840, where

$\frac{10,368}{2}$ = 5184 and $\frac{51,840}{10}$ = 5184 both of which relate directly to the "Atomic Weight of all 92 Natural Elements"

Taking the Trinity Function of 299,792,458, we get a terminal value equal to 136. The number 136 can be derived by adding (1 + 2 + 3 + 4 + 5 + 6 + 7 + 8 + 9 + 10 + 11 + 12 + 13 + 14 + 15 + 16) = 136. In a similar fashion the product of these numbers (1 x 2 x 3 x 4 x 5 x 6 x 7 x 8 x 9 x 10 x 11 x 12 x 13 x 14 x 15 x 16) = 20,922,789,888,000 = 5184 x 4,036,032,000 = 10368 x 2,018,016,000.

$$\text{"Speed of Light" squared} = C^2 = [299,792,458] \; times \; [299,792,458] = 8.9875517873681764 \times 10^{16}$$

If we take: 8.9875517873681764e+16 divided by 5184 =

0.001733709835526268595679012345 6790......(repeating digits out to infinity!)

Notice that the product of the repeating digits of 123456790 = 45,360 = 8.75 x 5184.

It appears to me that the number 5184 and 10368 must figure into Einstein's Law of Relativity:

$E = mC^2$. The numerical pattern of repeating digits above may well be related to the "particles" or the "waves" within the light.

The Relationship between the "Lord Jesus Christ" and the "Speed of Light" And the "Area of a Scalene Triangle"

A Scalene triangle has three (3) unequal sides. Let us use the numerical values for "Lord Jesus Christ" using Greek numerics, where "Lord" = 800 and "Jesus" = 888 and "Christ" = 1480.

Note: 800 + 888 + 1480 = 3168. Now let each of these number values represent the three (3) unequal sides, and let S equal the semi-perimeter of the triangle and let A equal the area of the scalene triangle.

Then Heron's formula is:

$A = \sqrt{S(S-800)(S-888)(S-1480)}$, where S = ½ (800 + 888 + 1480)

= 1584

$A = \sqrt{1584(1584-800)(1584-888)(1584-1480)}$

$A = \sqrt{89,890,504.704} = 299,817,452.30056 \times 10^{-3}$

Now the "Speed of Light" = 299,792,458 meters/second

Divide the "Speed of Light" by the Area of this Scalene triangle:

$\frac{299,792,458}{299,817,452} = 0.99991663594019\%$. Thus, the Area of this Scalene triangle is an extremely accurate value for the "Speed of Light".

Multiply the digits of 299,792,458 and we get: 2 x 9 x 9 x 7 x 9 x 2 x 4 x 5 x 2 = 3,265,920, where $\frac{3,265,920}{315} = 10,368$ and

$\frac{10,368}{2} = 5184$. 315 is just one of six permutations of 153, which is the "image" of Jesus in Greek numerics and 5184 is related directly to the "Atomic weights of the 92 natural elements".

"For I exist as …………Self-Generating Power"

Isaiah 46: 9 – 10 says, "Remember the former things of old: for I am God, and there is none else; I am God, and there is none like me, Declaring the end from the beginning, and from ancient times the things that are not yet done, saying, 'My counsel shall stand, and I will do all my pleasure'".

The numeric value of "self-generating power" is: Yod (10) + Beth (2) + Num (50) + Aleph (1) + Yod (10) + Beth (2) + Lamed (30) + Aleph (1) = 106 by addition, where the six (6) permutations of 106 are:

$106 + 160 + 61 + 16 + 610 + 601 = 1554$, where $(1554)^2 = 2,414,916$. When we multiply the digits of 2,414,916, we get: 2 x 4 x 1 x 4 x 9 x 1 x 6 = 1728, where 3 x 1728 = 5184 and 6 x 1728 = 10368.

Thus, we see that the "Self-Generating Power" is equivalent to "I am God", which is the "Lord Jesus Christ"!

Please see Dennis Lee Oberholtzer's book "Theophysic Gematria" listed in the bibliography for more details.

The "Lord Jesus Christ" is the "Aleph Tav" and the "Alpha Omega" Applying English numerics

"Aleph" = 6 + 72 + 30 + 96 + 48   and   "Tav" = 120+ 6 + 132

Multiplying the digits of Aleph Tav, we get: 6 x 7 x 2 x 3 x 9 x 6 x 4 x 8 x 1 x 2 x 6 x 1 x 3 x 2 =

31,352,832, where $\frac{31,352,832}{6048}$ = 5184, where 51,840 divided by 5 = 10368, which is just one of 120 permutations of 03168, the "Lord Jesus Christ" in Greek numerics.

"Alpha" = 6 + 72 + 96 + 48 + 6 and "Omega" = 90 + 78 + 30 + 42 + 6

Multiplying the digits of Alpha Omega, we get: 6 x 7 x 2 x 9 x 6 x 4 x 8 x 6 x 9 x 7 x 8 x 3 x 4 x 2 x 6 = 63,207,309,312, where $\frac{63,207,309,312}{12,192,768}$ = 5184, where 51,840 divided by 5 = 10368, which is just one of 120 permutations of 03168, the "Lord Jesus Christ" in Greek numerics.

If we add the Aleph Tav value of 31,352,832 and the Alpha Omega value of 63,207,309,312, we get the value of 63,238,662,144, where $\frac{63,238,662,144}{12,198,816}$ = 5184, where 51,480 divided by 5 = 10368, which is just one of 120 of permutations of 03168, the "Lord Jesus Christ" in Greek numerics. And the Trinity Function (63,238,662,144) = 153, which is the "image" of "Jesus" in Greek numerics. Note: These relationships are also the same if we subtract the value of Aleph Tav from the value of Alpha Omega and apply the same calculations to their difference in value.

Please note that these calculations for Aleph Tav and Alpha Omega give the same final result as found in the "Total Atomic Weights of all 92 Natural Elements". The number 5184 is found to be a common factor.

The Relationship between "Pi" and the "Speed of Light" and the "Lord Jesus Christ"

Pi = 3.14159265358

Speed of Light (c) = 299792458 meters/second

Digits of Pi = 3 x 1 x 4 x 1 x 5 x 9 x 2 x 6 x 5 x 3 x 5 x 8 = 3888000 where $\frac{3888000}{27000}$ = 144 and 144 x 36 = 5184.

Digits of (c) = 2 x 9 x 9 x 7 x 9 x 2 x 4 x 5 x 8 = 3265920 where $\frac{3265920}{22680}$ = 144 and 144 x 36 = 5184.

Thus, the "Digits of Pi" and the "Speed of Light" are both a function of number 144 and 5184.

"Pi" is the relationship between the diameter and the circumference of a circle.

This relationship (Pi) travels around the circle as a function of the speed of light (c).

Divide the digits of Pi by the reverse digits of Pi

$\frac{3.14159265358}{853562951413}$ = 0.000000000003681 = 3681^{-11}, 3681 is one permutation of 3168 = "Lord Jesus Christ" in Greek numerics. There are a total of 24 permutations of this number.

"God is Light, and in him is no darkness at all". I John 1:5

"It is He that sitteth upon the circle of the earth, and the inhabitants thereof are as grasshoppers". Isaiah 40:22

"For the eyes of the LORD run to and fro throughout the whole earth, to shew Himself strong in behalf of them whose heart is perfect toward Him". II Chronicles 16:9

"For His eyes are upon the ways of man, and He seeth his entire goings". Job 34:21

According to www.sciencenetlinks.com, earth's diameter = 12,756 kilometers, which equals 12,756,000 meters. 12,756 = 12 x 1063 (Prime #)

Strong's Number (Greek 1097) found in Revelation 3:3 has the meaning of "know", which has a numerics value of 1063.

Strong's Number (Hebrew 1363) found in 2 Chronicles 32:26 has the meaning of "himself for the pride" (Hebrew word "gobahh"), which has a numerics value of 12.

Since the diameter of the earth is 12,756,000 meters and since light speed travels at 299,792,000 meters/second, then $\frac{299,792,458}{12,756,000}$ = 23.50 seconds, which is the amount of time it takes light to travel one time around the earth in the area of "three hundred and sixty (360) degrees. But God is not limited by the "speed of light", which He created.

The "Surface of a Sphere" is $S=4\pi r^2$, where r = 6,378,000 meters. Then, S = 511,185,932,522,530 spherical meters. Since God sees the entire planet earth sphere continuously, then "God's speed of light" must be 1,705,132.730599 times faster than the "created speed of light = 299,792,458 meters/second".

Simply divide the "Surface of the Sphere" by the "created speed of light".

$$\frac{S\ meters}{C\ meters/second} = \frac{511{,}185{,}932{,}522{,}530}{299{,}792{,}458} = 1{,}705{,}132.730599.$$ We can see that the "created speed of light" is only a small fraction of "God's speed of light":

$$\frac{299{,}792{,}458}{511{,}185{,}932{,}522{,}530} = 0.000000586464609.$$

Relationship between Euler's e and the "Lord Jesus Christ"

We remove the decimal point in Euler's e and use the first 13 digit of this mathematical constant: 2718281828459

Multiplying the digits we get: 2 x 7 x 1 x 8 x 2 x 8 x 1 x 8 x 2 x 8 x 4 x 5 x 9 = 41,287,680

Now the Trinity Function (2718281828459) = 371, which is just one of 6 permutations of 137, the approximate inverse of the Fine Structure Constant.

Also, the six permutations of 371 are 371 + 317 + 137 + 173 + 713 + 731 = 2442

Now the Trinity Function (41,287,680) = 153 = the "image" of Jesus in Greek numerics

Also, the six permutations of 153 are 153 + 135 + 315 + 351 + 513 + 531 = 1998

Now multiplying 1998 x 2442 = 4,879,116 = 36 x 135,531 = 37 x 131,868 = 148.5 x 888, where 888 = "Jesus" in Greek numerics. Also, 135 and 531 are just 2 of the permutations of 153 = the "image" of Jesus in Greek numerics.

Also, 100 x 148.5 = 14850, which is just one of 120 permutations of 15840, where $\frac{15840}{5}$ = 3168 = "Lord Jesus Christ" in Greek numerics. Also, 14,850 is just one of 120 permutations of 51840, where 5184 is one of the mathematical relationships with the "Total Atomic Weights of the 92 Natural Elements".

Characteristics of the "Fine Structure Constant"

CODATA Value – 2006 for the Fine Structure Constant is 0.0072973525376 from Physics Today.

If we sum the cube of each digit in this number, we get the following value: $7^3 + 2^3 + 9^3 + 7^3 + 3^3 + 5^3 + 2^3 + 5^3 + 3^3 + 7^3 + 6^3 = 2294$.

The Trinity Function (72973525376) = 407 = [370 + 37] = Hebrew numeric value for "and" (which stands between the "heavens" and the "earth") in Genesis 1:1. Also, 407 = "The circuit of heaven" in Hebrew numeric (see Job 22:14).

Note that 6438 – 2368 = 4070, where 2368 = "Jesus Christ" in Greek numeric and 4070 is just one of 24 permutations of 0407 or 407. Note that 6438 is derived by the following transactions 371 -137 = 234 and 407 -137 = 270 and 407 -370 = 37 and 407 – 371 = 36 and 153 + 217 = 370

2442 = Sum of all permutations of 137 and 1998 = Sum of all permutations of 153

And 1998 = Sum of all permutations of 234. Now all 2442 + 1998 + 1998 = 6438

Also, 7 x 2 x 9 x 7 x 3 x 5 x 2 x 5 x 3 x 7 x 6 = 166698. Take 166698 + 896661 = 729963, which shows two connections to "Jesus Christ": 729963 = 153 x 4771, where 153 is the "image" of "Jesus" in Greek numeric when the Trinity Function is applied to 888 = "Jesus"; and 729963 = 117 x 6239, where 1 + 6238 = 6239 and 6238 is just one of 24 permutations of "Jesus Christ" = 2368 in Greek numerics.

Take the first 4 digits of alpha α = 7297 and apply the Trinity Function (7297) = 1. Also, the Sum of all permutations of 7297 = 166650 and 1 x 6 x 6 x 6 x 5 = 1080 = "God himself who formed the earth and made it" in Hebrew numerics value (see Isaiah 45:18). If we multiply the digits of the Fine Structure Constant = 72,973,525,376, we get these results: 7 x 2 x 9 x 7 x 3 x 5 x 2 x 5 x 3 x 7 x 6 = 16,669,800.

If we multiply the digits of 16, 669,800, we get these results: 1 x 6 x 6 x 6 x 9 x 8 = 15,552, where $\frac{15,552}{3}$ = 5184, where 51,840 divided by 5 = 10368, which is just one of 120 permutations of 03168, which is the "Lord Jesus Christ" in Greek numerics.

Also, if we apply the Trinity Function (5184) = 371, which is just one of six permutations of 137, the approximate inverse of the Fine Structure Constant.

Also, if we take the value for the multiplied digits of the Fine Structure Constant, which equals 16,669,800 and calculate $\frac{166,698}{27}$ = 6174.

Then we find the 24 total permutations of 6174 = 119,988, where the product of these digits 1 x 1 x 9 x 9 x 8 x 8 = 5184. Note that the Hebrew numerics value for "light" is equal to 207 or 27 by dropping the zero.

Thus, we can see that the number 27 and the number 5184 play a major role in the Fine Structure Constant.

Thus, when we multiply the digits of the Fine Structure Constant = 729735253765, we get these results where 5184 relates directly to the "Atomic Weights of the 92 Natural Elements".

The Atomic Weight of Gold and its relationship with the "Lord Jesus Christ"

The Atomic Weight of Gold = 1,969,665,694

If we multiply the digits of 1,969,665,694, we get: 1 x 9 x 6 x 9 x 6 x 6 x 5 x 6 x 9 x 4 = 18,895,680, where $\frac{18,895,680}{3645}$ = 5184, which is the number which links us to 51,840 divided by 5 = 10368, which is one of 120 permutations of 03168 = the "Lord Jesus Christ" in Greek numerics.

Also, if we find all 24 permutations of 3645, we get the sum total = 119,988, where 1 x 1 x 9 x 9 x 8 x 8 = 5184, which is the same sum total for the 24 permutations of 6174 found in the Fine Structure Constant.

Please note that I randomly chose to calculate these values for the GOLD element; but I suspect that each of the 92 natural elements have the same characteristics.

The "Golden Ratio" (Phi ϕ) and the "Lord Jesus Christ"

Phi (ϕ) = 1618033988749 without any decimal point

The Trinity Function (1618033988749) = 136, where $1^3 + 3^3 + 6^3$ = 244 and $2^3 + 4^3 + 4^3 = 136$

If we multiply the digits of Phi (ϕ), we get: 1 x 6 x 1 x 8 x 3 x 3 x 9 x 8 x 8 x 7 x 4 x 9 = 62,705,664

The Trinity Function (62,705,664) = 153, which is the "image" of Jesus in Greek numerics and $\frac{62,705,664}{12,096}$ − 5184, where 5184 relates directly to the "Total Atomic Weights of the 92 Natural Elements". Also, 1584 is just one of 24 permutations of 5184, where 2 x 1584 = 3168, which is the Greek numerics value for the "Lord Jesus Christ".

Notice too that if we multiply the digits of 62,705,664, we get: 6 x 2 x 7 x 5 x 6 x 6 x 4 = 60,480, where $\frac{60,480}{5}$ = 12,096. Also 60,480 is just one of 120 permutations of 06048. Now taking the 24 permutations of 6048, we get the following values:

6048 + 6084 + 6804 + 6840 + 6408 + 6480 = 38,664 and 0648 + 0684 + 0468 + 0486 + 0846 + 0864 = 3,996 and 4068 + 4086 + 4608 + 4680 + 4806 + 4860 = 27,108 and 8046 + 8064 + 8406 + 8460 + 8604 + 8640 = 50,220.

Then, 38,664 + 3,996 + 27,108 + 50,220 = 119988, where 1 x 1 x 9 x 9 x 8 x 8 = 5184, which again relates directly to the "Total Atomic Weights of the 92 Natural Elements".

If we add Phi (φ) = 1,618,033,988,749 and 9,478,893,308,161 Reverse Phi (φ), we get this value: 11,096,927,296,910. Now by multiplying the digits of 11,096,927,296,910, we get a value = 6,613,488.

Then, 1458 x 4536 = 6,613,488, where is 1458 is just one of 24 permutations of 5184, which again relates directly to "Total Atomic Weights of the 92 Natural Elements".

Also, notice that the Trinity Function (11,096,927,296,910) = 371, where 371 is just one of six permutations of 137, which is the approximate value of $\frac{1}{137}$ = Fine Structure Constant.

IUPAC Commission on Atomic Weights and Isotropic Abundances

ATOMIC WEIGHTS OF THE ELEMENTS 2007

1   H   Hydrogen   1.00794(7)
2   He   Helium   4.002602(2)
3   Li   Lithium   6.941(2)
4   Be   Beryllium   9.012182(3)
5   B   Boron   10.811(7)
6   C   Carbon   12.0107(8)
7   N   Nitrogen   14.0067(2)
8   O   Oxygen   15.9994(3)
9   F   Fluorine   18.9984032(5)
10 Ne   Neon   20.1797(6)
11 Na   Sodium   22.98976928(2)
12 Mg   Magnesium   24.3050(6)
13 Al   Aluminum   26.9815386(8)
14 Si   Silicon   28.0855(3)
15 P   Phosphorus   30.973762(2)
16 S   Sulfur   32.065(5)
17 Cl   Chlorine   35.453(2)
18 Ar   Argon   39.948(1)
19 K   Potassium   39.0983(1)
20 Ca   Calcium   40.078(4)
21 Sc   Scandium   44.955912(6)

22	Ti	Titanium	47.867(1)
23	V	Vanadium	50.9415(1)
24	Cr	Chromium	51.9961(6)
25	Mn	Manganese	54.938045(5)
26	Fe	Iron	55.845(2)
27	Co	Cobalt	58.933195(5)
28	Ni	Nickel	58.6934(2)
29	Cu	Copper	63.546(3)
30	Zn	Zinc	65.409(4)
31	Ga	Gallium	69.723(1)
32	Ge	Germanium	72.64(1)
33	As	Arsenic	74.92160(2)
34	Se	Selenium	78.96(3)
35	Br	Bromine	79.904(1)
36	Kr	Krypton	83.798(2)
37	Rb	Rubidium	85.4678(3)
38	Sr	Strontium	87.62(1)
39	Y	Yttrium	88.90585(2)
40	Zr	Zirconium	91.224(2)
41	Nb	Niobium	92.90638(2)
42	Mo	Molybdenum	95.94(2)
43	Tc	Technetium	[98]
44	Ru	Ruthenium	101.07(2)
45	Rh	Rhodium	102.90550(2)
46	Pd	Palladium	106.42(1)
47	Ag	Silver	107.8682(2)
48	Cd	Cadmium	112.411(8)
49	In	Indium	114.818(3)
50	Sn	Tin	118.710(7)
51	Sb	Antimony	121.760(1)
52	Te	Tellurium	127.60(3)
53	I	Iodine	126.90447(3)
54	Xe	Xenon	131.293(6)
55	Cs	Caesium	132.9054519(2)
56	Ba	Barium	137.327(7)
57	La	Lanthanum	138.90547(7)
58	Ce	Cerium	140.116(1)
59	Pr	Praseodymium	140.90765(2)
60	Nd	Neodymium	144.242(3)
61	Pm	Promethium	[145] 5
62	Sm	Samarium	150.36(2)
63	Eu	Europium	151.964(1)
64	Gd	Gadolinium	157.25(3)

65	Tb	Terbium	158.92535(2)
66	Dy	Dysprosium	162.500(1)
67	Ho	Holmium	164.93032(2)
68	Er	Erbium	167.259(3)
69	Tm	Thulium	168.93421(2)
70	Yb	Ytterbium	173.04(3)
71	Lu	Lutetium	174.967(1)
72	Hf	Hafnium	178.49(2)
73	Ta	Tantalum	180.94788(2)
74	W	Tungsten	183.84(1)
75	Re	Rhenium	186.207(1)
76	Os	Osmium	190.23(3)
77	Ir	Iridium	192.217(3)
78	Pt	Platinum	195.084(9)
79	Au	Gold	196.966569(4)
80	Hg	Mercury	200.59(2)
81	Tl	Thallium	204.3833(2)
82	Pb	Lead	207.2(1)
83	Bi	Bismuth	208.98040(1)
84	Po	Polonium	[209]
85	At	Astatine	[210]
86	Rn	Radon	[222] 5
87	Fr	Francium	[223]
88	Ra	Radium	[226] 5
89	Ac	Actinium	[227]
90	Th	Thorium	232.03806(2)
91	Pa	Protactinium	231.03588(2)
92	U	Uranium	238.02891(3)

Total Atomic Weight of all Natural Elements = 10,310.6645966

Please note these calculations:
$[1 \times 3 \times 1 \times 6 \times 6 \times 4 \times 5 \times 9 \times 6 \times 6] = 699840 = 648 \times 1080$, where $648 = 3 \times 6^3 = 2 \times 18^2$ and $1080 = 27 \times 40$ and $48 \times 1080 = 51840$ and $96 \times 1080 = 10368$ and $16 \times 648 = 10368$ and $8 \times 648 = 5184$. Note: $699840 = 135 \times 5184 = 67,500 \times 10.368$, which mathematically relates to the Lord Jesus Christ (3168) and "that certain" (2880) in Daniel 8:13. Note that 1584 is one of 24 permutations of 5184 and $1584 = \frac{3168}{2}$.

It appears to me that the numbers 5184 and 10368 and 3168 and 7 are directly related to the "Total Atomic Weights of the 92 Natural Elements", as shown below:

103106645966 divided by 5184 =

19889399.2989969135802469..........

The series of digits 135802469 repeat continuously out to infinity!

Notice the repeating digits of 135802469 above. When these digits are multiplied together, we get a value of 51,840.

103106645966 divided by 3168 =

32546289.76199494949494.........

The series of digits 9494949494 repeat continuously out to infinity!

Notice the repeating digits of 9494949494 above. When these digits are multiplied together, you get a total value = 6718464 = 1296 x 5184.

103106645966 divided by 7 =14729520852.285714...........

Notice that the series of digits 285714 are continuously repeated out to infinity! Notice that when number 7 is divided into the "Total Atomic Weights of the 92 Natural Elements", we get repeating digits of 285714, which is the same as when we divide 7 into number 2. Trinity Function [699840] = 153, which is the "image" of Jesus (888 in Greek numerics).

Prime Factors of 103,106,645,966 = 2 x 419 x 123,038,957. The Trinity Function [103106645966] = 371

Please note that 648 is just one of 6 permutations of 864. Please note that 16 x 648 = 10368 and 80 x 648 = 51840. The numbers 864 and 1080 and have special Biblical meanings, as listed below:

864 = "God" in Greek numerics

864 = "Cornerstone" in Greek numerics

864 = "God of Peace" in Greek numerics

864 = "God is fire" in Greek numerics (see Hebrews 12:29)

864 = "Life" in Greek numerics

864 = "The word of the LORD from Jerusalem" in Hebrew numerics (see Micah 4:2)

864 = "Before me there was no God formed, neither shall there be after me" in Hebrew numerics (see Isaiah 43:10)

1080 = "God himself who formed the earth and made it" in Hebrew numerics (see Isaiah 45:18)

1080 = "The Holy Spirit" in Greek numerics

1080 = "Heaven is my throne and the earth is my footstool" in Hebrew numerics (see Isaiah 66:1)

370 = "My Messiah" in Hebrew numerics (see Psalm 105:15)

370 = "He has founded the earth" in Hebrew numerics

370 = "He lives" in Hebrew numerics

370 = "He rules" in Hebrew numerics (see Psalm 66:7)

Note: The above scriptural references and numerics values were calculated by Bonnie Gaunt. Please see her books listed in the bibliography. Notice that 96 x 108.0 = 10368 and 480 x 108.0 = 51840.

Relationships between Physical and Mathematical Constants

In this section I will show the many relationships among the physical and mathematical constants. It appears that all physical and mathematical constants are inter-connected when you can find the appropriate "root" relationship.

This list is by no means comprehensive or complete; there are many other physical and mathematical relationships that should be analyzed. Please do not think that other significant relationships do not exist, because I am convinced that they do. Notice how the numbers 5184 and 10368 can also be found within these relationships if the necessary calculations are performed. Some these constants, such as Pi or Euler's e or the Golden Ratio ϕ, are transcendental and irrational; therefore, their "roots" are also transcendental and irrational.

Notice too that for the purposes of analyzing these constants, I have ignored the decimal points which are simply place holders and I have also ignored the systems of measurement.

I am only looking at the raw digits for each of these constants. To completely analyze these many relationships, a super computer will be necessary to find the appropriate "root" as it approaches infinity.

- Relationship between Planck's Constant and π
  Planck's constant = $6.62606896 \times 10^{-34}$ joule-seconds

  $$\sqrt[17.74365527]{662606896} = \pi = 3.14159265$$

- Relationship between the Speed of Light in a Vacuum and π
  The Speed of Light in a Vacuum = 299792458 m/s^{-1}

  $$\sqrt[17.05083551]{299792458} = \pi = 3.14159265$$

- Relationship between the Elementary Electron Charge and π
  Elementary Electron Charge = $1.602176487 \times 10^{-19}$

  $$\sqrt[18.5149607]{1602176487} = \pi = 3.141592653$$

- Relationship between Alpha (α) (The Fine Structure Constant) and the Number 37
  Alpha (α) = 0.007297351

  $$\left[\sqrt[496.6291]{37}\right] - [1] = 0.007297351$$
  Note: 7297351 is Prime Number

  Relationship between the Inverse of Alpha (α^{-1}) and Euler's e
  Inverse of Alpha (α^{-1}) = 137.035999111 and e = 2.718281828459

  $$\sqrt[137.035999111]{2.718281828459} = 1.007324043*$$

40

- Relationship between the Inverse of Alpha (α^{-1}) and π
  Inverse of Alpha (α^{-1}) = 137.035999111 and π = 3.1415926535898
  $$\sqrt[137.035999111]{3.1415926535898} = 1.008388485*$$
  *Note that 1.007324043 + 1.008388485 = 2.105712528
  Then $\frac{2.105712528}{2}$ = 1.007856264, which is just 0.000031232 different from the "Relative Atomic Mass" of Hydrogen (H) according to the "National Institute of Standards and Technology" value.

  Note that the Trinity Function (72973525678) goes to 2779, goes to 1423, goes to 100, goes to 1, where 2779 is a permutation of 7297.

- Relationship between Phi (φ) (Also called the Golden Ratio or the Divine Proportion) and π
  Phi (φ) = 1.6180339887 and π = 3.1415926535898

  $$\sqrt[2.37884821]{3.1415926535898} = \text{Phi (φ)} = 1.618033987$$

- Relationship between Euler's e and π
  Euler's e = 2.71828182884 and π = 3.1415926535898

  $$\left[\sqrt[3.73055033]{3.1415926535898}\right] \times [2] = 2.71828182884$$

- Relationship between Alpha (α) and π
  Alpha (α) = 0.007297351 and π = 3.1415926535898

  $$\left[\sqrt[157.4409]{3.1415926535898}\right] - [1] = 0.007297351$$

- Relationship between the Number 37 and π
  Number 37 is the 13th prime number (where 1 is counted as prime) $\left[\sqrt[78.2981]{37}\right] \times [3] = 3.1415926$
  Note: 782981 is the 62,722th prime number

- Relationship between Euler's e = 2.71828182884 and π = 3.1415926535 and Phi (φ) = 1.6180339887

  $2.71828182884 + 3.1415926535 + 1.6180339887 =$
  $7.477908469 \left[\sqrt[43.62658]{7.477908469}\right] \times [3] =$
  $3.14159265 = \pi$

  Please note that 314159 and 951413 are prime numbers and 1618033 and 3308161 are prime numbers and 2718281 is a prime number.

  Also, multiply the digits of 7,477,908,469: 7 x 4 x 7 x 7 x 9 x 8 x 4 x 6 x 9 = 21,337,344, where
  2,058 x 10,368 = 21,337,344 and 4,116 x 5,184 = 21,337,344.

  Also, multiply the digits of 271828182884: 2 x 7 x 1 x 8 x 2 x 8 x 1 x 8 x 2 x 8 x 8 x 4 = 7,340,032 and 7,340,032 + 2,300,437 = 9,640,469, where 9 x 6 x 4 x 4 x 6 x 9 = 46,656 = 9 x 5184 and 466,560 = 45 x 10,368.
  Also, multiply the digits of 31415926535: 3 x 1 x 4 x 1 x 5 x 9 x 2 x 6 x 5 x 3 x 5 = 486,000, where 9375 x 5184 = 48,600,000 and 46,875 x 10368 = 486,000,000

Also, multiply the digits of 16180339887: 1 x 6 x 1 x 8 x 3 x 3 x 9 x 8 x 8 x 7 = 1,741,824, where 336 x 5184 = 1,741,824 and 168 x 10,368 = 1,741,824.

- Relationship between the first 10 natural numbers (1,2,3,4,5,6,7,8,9,10) and π and Euler's e

    Take the prime number 1234567891 and find the roots for π and Euler's e
    $\sqrt[18.28727206]{1234567891} = \pi = 3.141592654$ and
    $\sqrt[20.93398686]{1234567891} =$ Euler's e

    Also, multiply the digits of 12345678910: 1 x 2 x 3 x 4 x 5 x 6 x 7 x 8 x 9 x 1 = 362,880, where
    70 x 5184 = 362,880 and 35 x 10,368 = 362,880.

- Relationship between the "Inverse of Pi", the "Inverse of e", the "Inverse of Phi" and Alpha(α)
    $\frac{1}{\pi} = 0.3183098861838$ and $\frac{1}{e} = 0.3678794411714$ and $\frac{1}{\phi} = 0.6180339887498$
    Then $\left[\frac{1}{\pi}\right] \times \left[\frac{1}{e}\right] \times \left[\frac{1}{\phi}\right] = 0.072371571835 \times 10^{-1}$
    Then Alpha (α) = 0.0072973525678 minus 0.0072371571835 = 0.0000601953541 difference

- Relationship between the "Atomic Weight of all 92 Natural Elements" and the "Inverse of Alpha"
    The atomic weight of all 92 natural elements = 10232.164593832
    Then $\frac{1.0232164593832 \times 10^{-4}}{173.035999111} = 0.0074667712573$, which differs from Alpha (α) by just 0.0001694186895

- Relationship between the "Number of Days in a Solar Year" and Alpha (α)

  The number of days in a solar year equals 365.2421896698

  Then $\frac{1}{365.2421896698} = 0.0027379093332$.

  Alpha (α) = 0.0072973525678 minus 0.0027379093332 = 0.0045594432345 difference from Alpha (α)

- Relationship between the "Inverse of Alpha (α^{-1})" with Euler's e and π and the "Relative Atomic Mass" of Hydrogen (H)

  "Inverse of Alpha (α^{-1})" = 137.035999111

  Then, $\sqrt[137.035999111]{2.718281828459} = 1.007324043$ and $\sqrt[173.035999111]{3.141592653589} = 1.008388485$

  Then $\frac{1.007324043 + 1.008388485}{2} = 1.007856264$

"Relative Atomic Mass" of Hydrogen (H) = 1.0078250321, which differs only 0.000031232 from the average of the "Inverse of Alpha (α^{-1}) root of Pi and Euler's e".

- Relationship between Prime Numbers and π

  The first 10 Prime Numbers are 1,2,3,5,7,11,13,17,19,23. The write these Prime Numbers as one (1) large number: 123571113171923

  Note: 1235711131 and 1235711 are Prime Numbers

  $\sqrt[18.28808063]{1235711131} = \pi = 3.141592654$

  Also, multiply the digits of 123571113171923: 1 x 2 x 3 x 5 x 7 x 1 x 1 x 1 x 3 x 1 x 7 x 1 x 9 x 2 x 3 = 238,140. Then $(238,140)^2 = 56,710,659,600$.

Again, we multiply the digits: 5 x 6 x 7 x 1 x 6 x 5 x 9 x 6 = 340,200. Then $(340,200)^2$ = 115,736,040,000 = 5184 x 22,325,625 and 10,368 x 111,628,125 = 1,157,360,400,000

Choose another prime number randomly: 225871. $(225871)^2$ = 51,017,708,641, where we then multiply the digits: 5 x 1 x 1 x 7 x 7 x 8 x 6 x 4 x 1 = 47,040. And $(47,040)^2$ = 2,212,761,600, where we again multiply the digits: 2 x 2 x 1 x 2 x 7 x 6 x 1 x 6 = 2,016. And $(2,016)^2$ = 4,064,256 = 784 x 5184 and 392 x 10,368 = 4,064,256.

Choose another prime number randomly: 626389. Multiply the digits, we get: 6 x 2 x 6 x 3 x 8 x 9 = 15,552, where 3 x 5184 = 15,552 and 15 x 10,368 = 155,520.

Choose another prime number randomly: 1076513. $(1076513)^2$ = 1,158,880,239,169, where we then multiply the digits: 1 x 1 x 5 x 8 x 8 x 8 x 2 x 3 x 9 x 1 x 6 x 9 = 7,464,960 = 1440 x 5184 and 720 x 10368.

Choose another prime number randomly: 48112959837082048697. When we multiply the digits, we get: 4 x 8 x 1 x 1 x 2 x 9 x 5 x 9 x 8 x 3 x 7 x 8 x 2 x 4 x 8 x 6 x 9 x 7 = 842,764,124,160.
51840 x 16,257,024 = 842,764,124,160 and 10,368 x 81,285,120 = 842,764,124,160.

- Relationship between the "Relative Atomic Mass" of Hydrogen (H) and π :

- "Relative Atomic Mass" of Hydrogen (H) = 1.0078250321.
  Then $\sqrt[146.86239]{3.141592653589}$ = 1.007825032

- Relationship between the "Standard Atomic Weight" of Hydrogen (H) and π
  "Standard Atomic Weight" of Hydrogen (H) = 1.00794(7)
  Then $\sqrt[144.615]{3.141592653589}$ = 1.007947118

- Relationship between the 40th Fibonacci number (102334155) and π
  Then $\sqrt[947.76185]{31415926535}$ x 10^8 = 102334155

- Relationship between God's Numbers and π = 3.1415926535 and Phi (ϕ) = 1.6180339887
  God's Numbers: 3, 7, 10, 12, 21, 22, 37. Write these numbers as one large natural number: 371012212237; then 732212210173 is a Prime Number! Now multiply these individual numbers: 3 x 7 x 10 x 12 x 21 x 22 x 37 = 43076880
  Then $\left[\sqrt[370]{43076880}\right]$ x [3] = 3.145968337, which varies from π by just 0.004375684 difference. Note: $\frac{22}{7}$ = 3.142857143 is a commonly used value for π.
  Then $\left[\sqrt[37]{43076880}\right]$ = 1.608166183, which varies from Phi (ϕ) by just 0.009867805 difference.
  Then 371012212237 x 732212210173 = 2.716596719322, which varies just 0.0016851095180 from the value for Euler's e.

- Relationship between π, e, ϕ and the Area of a Scalene Triangle

  A scalene triangle has three (3) unequal sides. To find the area of this triangle we use Heron's Formula: Let π, e, ϕ be the three (3) unequal sides and let S equal the semi-perimeter of the triangle and let A equal the area of the scalene triangle.

  Then Heron's Formula is:
  $A = \sqrt{S(S-\pi)(S-e)(S-\phi)}$, where S = ½ (π + e + φ).
  Then,
  π= 3.1415926535 + e = 2.7182818284 + 1.6180339887 = 7.477908469. Then, S = ½ (π + e + φ) = ½ (7.477908469) = 3.738954235, which is the semi-perimeter of the triangle. Then, the area (A) of the triangle equals:

  $\sqrt{(3.738954235)(3.738954235 - 3.1415926535)(3.738954235 - 2.7182818284)(3.738954235 - 1.6180339887)} = \sqrt{4.83501861} =$ 2.198867574 = A the area of a scalene triangle where the three (3) unequal sides are π, e, φ

  Note that 219886757 is a Prime Number!

Prime Numbers are directly related to the "Total Atomic Weights of the 92 Natural Elements"

A scalene triangle has three (3) unequal sides. To find the area of this triangle, we use Heron's Formula:

$$A = \sqrt{S(S-x)(S-y)(S-z)}, \text{ where } S = \frac{1}{2}(x + y + z)$$

Let x and y and z be the three (3) unequal sides and let S equal the semi-perimeter of the triangle and let A equal the area of the scalene triangle.

Randomly choose 3 prime numbers to represent the three (3) unequal sides, where x = 9011 and y = 3769 and z = 7499 and S = $\frac{1}{2}$(9011 + 3769 + 7499) = 10,139.5.

Then, A =
$$\sqrt{10,139.5(10,139.5 - 9011)(10,139.5 - 33769)(10,139.5 - 7499)}$$
or

A = 13,873,591.328175. Then by multiplying the digits of A, we get a value of 38,102,400, where

38,102,400 = 5184 x 7350. Note that 5184 is directly related to the "Total Atomic Weights of the 92 Natural Elements" and 3675 x 10368 = 38,102,400.

Randomly choose 3 prime numbers to represent the three (3) unequal sides, where x = 5591 and y = 3593 and z = 8389 and S = $\frac{1}{2}$(5591 + 3593 + 8389) = 8,786.5.

Then, A =

$$\sqrt{8,786.5(8,786.5 - 5591)(8,786.5 - 3593)(8,786.5 - 8389)}$$

or A = 7,613,353.6203034. Then by multiplying the digits of A, we get a value of 2,449,440, where

24,494,400 = 5184 x 4725. Note that 5184 is directly related to the "Total Atomic Weights of the 92 Natural Elements" and 23,625 x 1036.8 = 24,494,400.

Randomly choose 3 prime numbers to represent the three (3) unequal sides, where x = 7013 and y = 4931 and z = 2017 and S = $\frac{1}{2}$(7013 + 4931 + 2017) = 6980.5

Then, A =

$$\sqrt{6980.5(6980.5 - 7013)(6980.5 - 4931)(6980.5 - 2017)}$$

or

A = 1,519,157.9147764. Then by multiplying the digits of A, we get a value of 66,679,200, where

66,679,200 = 103.68 x 643,125. Note that 6,667,920,000 = 10368 x 643,125, where 10368 is just one of 120 permutations of 03168, which is 3168 = "Lord Jesus Christ" in Greek numerics and $\frac{10368}{2}$ = 5184, which is directly related to the "Total Atomic Weights of the 92 Natural Elements". Also, note in the above equation that (6980.5 - 7013) would produce a negative value of -32.5, so I changed the sign to +32.5 to make the equation calculation possible. This sign change does not seem to affect the final results.

The prime numbers used in the above 3 examples were randomly chosen so that this could possibly show that all prime numbers have a "triangular relationship" with the "Total Atomic Weights of the 92 Natural Elements" and the "Lord Jesus Christ" in Greek numerics. Since many prime numbers are extremely large, it will probably be necessary to use mainframe computers to further test these "triangular relationships". I am convinced that there are an infinite number of triangular relationships among all of the prime numbers. These triangular relationships are a definite "pattern" used in construction of the prime numbers!!

Fibonacci Numbers

The first 12 Fibonacci Numbers are:

1, 1, 2, 3, 5, 8, 13, 21, 34, 55, 89, and 144

$1 + 1 + 2 + 3 + 5 + 8 + 13 + 21 + 34 + 55 + 89 + 144 = 376$

$1 \times 1 \times 2 \times 3 \times 5 \times 8 \times 13 \times 21 \times 34 \times 55 \times 89 \times 144 = 1{,}570{,}247{,}078{,}400$

Then $\frac{1{,}570{,}247{,}078{,}400}{151{,}451{,}300} = 10368$ and $\frac{1{,}570{,}247{,}078{,}400}{30{,}290{,}260} = 51840$. Notice that 3168 = Greek numerics value of "Lord Jesus Christ" and 51840 relates directly to the "Total Atomic Weights of the 92 Natural Elements".

The Trinity Function of 1,570,247,078,400 = 153, where 1,570,247,078,400 is the "pre-image" and 153 is the "image" of Jesus in Greek numerics.

The Trinity Function of 376 = 370. All the permutations of 376 are:

376 + 367 + 763 + 736 + 673 + 637 = 3552 = 4 x 888, where 888 is the Greek numerics value of "Jesus".

Kaprekar Constants and Numbers in relationship to Number 7 and Number 31415926 = Pi

Kaprekar Constant: Take any number whose digits are not all identical. Rearrange the digits to make the largest and the smallest numbers possible. Subtract the smaller number from the larger number. Use the resulting number and repeat the process until it results in a non-zero constant, zero, or a cycle.

Number 7. All the divisions of 7 into 1, 2, 3, 4, 5, 6 produce a permutation of 142857. Taking this number, let's apply the Kaprekar Constant process.

875421 − 124578 = 750843

875430 − 34578 = 840852

885420 − 24588 − 860832

886320 − 23688 = 862632

866322 − 223668 = 642654

665442 − 244566 = 420876

876420 − 24678 = 851742

875421 − 124578 = 750843 Thus, the Kaprekar Constant finishes with a cyclical number.

Number 31415926

96543211 − 11234569 = 85308642

88654320 − 2345688 = 86308632 (the beginning of the cycle)

88663320 − 2336688 = 86326632

86663322 − 22336668 = 64326654

66654432 − 23445666 = 43208766

87664320 − 2346678 = 85317642

87654321 − 12345678 = 75308643

87654330 − 3345678 = 84308652

88654320 − 2345688 = 86308632 (the end of the cycle). The Kaprekar Constant appears to show that the digits of Pi are cyclical. Notice that 8632 is a permutation of 2368 = "Jesus Christ" in Greek numerics.

Kaprekar Number: Take any positive whole number n that has d number of digits. Then take the square of n and separate the result into two pieces: the right-hand piece that has d digits and the left hand piece has (d) or (d − 1) digits. Now add these two pieces together. If the result is n, then n is a Kaprekar Number.

Take the Number 7 and divide it into 1, 2, 3, 4, 5, and 6 respectively: $\frac{1}{7}$ = .142857 and $\frac{2}{7}$ = .285714 and $\frac{3}{7}$ = .428571 and $\frac{4}{7}$ = .571428 and $\frac{5}{7}$ = .714285 and $\frac{6}{7}$ = .857142

If we apply the Kaprekar Number formula to these decimal values (by first removing the decimal point), there is only one Kaprekar Number, which is 142857. The other numbers above are NOT Kaprekar Numbers, but they produce permutations of 142857 under this formula. If we apply the Trinity Function to 142857 we immediately get Number 153, which is the "image" of Jesus in Greek numerics.

If we take the decimal values of $\frac{x}{7}$ as listed above and remove the decimal points and then divide each number by 360° in a circle, we get the following values:

142857 produces	117° + 180° = 297°
285714 produces	54° + 180° = 234°
428571 produces	171° + 180° = 351°
571428 produces	108° + 180° = 288°
714285 produces	45° + 180° = 225°
857142 produces	162° + 180° = 342°
$\frac{7}{7} = 1$ produces	1° + 180° = 181°

When we add 142857 + 285714 + 428571 + 571428 + 714285 + 857142 + 1 = 2,999,998. The multiply the digits of 2,999,998 and we get 944,784 = 18,225 x 51.84 = 91,125 x 10.368, where 10368 is related to the "Lord Jesus Christ" = 3168 in Greek numerics and 5184 is related to the "Total Atomic Weights of the 92 Natural Elements".

When these degrees are plotted onto a circle, the diameters of the circle produce a six pointed star.

Notice the following characteristics of these various degrees numbers:

Number 45 is the smallest Kaprekar Number after 1 and 9; it is also the 5th hexagonal number.

Number 54 is the reverse of 45 and $\frac{1}{2}$ x 108 = 54 and 2 x 27 = 54

Number 108 is the smallest of the heptominoes and there are only 108 heptominoes. Also, $\frac{1}{2}(6^3)$ = 108

Number 117 = $\frac{1}{2}$(234), where 232, 233, 234 is the smallest triple of consecutive numbers each of which is the sum of 2 squares: 232 = $6^2 + 14^2$ and 233 = $8^2 + 13^2$ and 234 = $3^2 + 15^2$. Also, 3 x 117 = 351 and 9 x 117 = 1053, both of which are the Trinity Function "image value" of Jesus.

Number 162: 2 x 162 = 324 and 162° + 180°= 342° and 324 + 342 = 666

Number 171   2 x 171 = 342 and 3 x 171 = 513 and 171°+ 180° = 351°

Number 181 is the 43rd Prime Number

Number 225   5 x 45 = 225

Number 234   See Number 117 above.

Number 288   108° + 180° = 288° and $\frac{1}{2}$(288) =144

Number 297 is the 6th Kaprekar Number.

The first 12 Kaprekar Numbers are:

1, 9, 45, 55, 99, 297, 703, 999, 2223, 2728, 7272, 7777

Note that the smallest 10 digit Kaprekar Number is 1,111,111,111 whose square is 12345678900987654321

The Number 7 and the Speed of Light

The number seven (7) is directly related to the number 5184 and 10368 and the Speed of Light, as will be shown below.

$\frac{1}{7}$ = 142857 = 1.485 x 96,200 = 2.97 x 48,100, where 1485 is just one permutation of 5184 and 1485 x 2 = 2970. Notice that the number 142857 when rotated onto a circle of 360°, it stops on 297°. For example, $\frac{142857}{360}$ = 396.825. So 396 x 360 = 142560 and 142857 − 142560 = 297. You can use this procedure on the other number seven (7) calculations.

$\frac{2}{7}$ = 285714 = 1.584 x 180,375 = 0.3168 x 901,875, where 2 x 1.584 = 3.168. When the number 285714 is rotated on a circle, it stops at 234°.

$\frac{3}{7}$ = 428571 = 1.485 x 288,600 = 2.97 x 144,300, where 2 x 1.485 = 2.970. When the number 428571 is rotated on a circle, it stops at 351°, where 351 is just one of six (6) permutations of 153 = "Jesus" in Greek numerics.

$\frac{4}{7}$ = 571428 = 15.84 x 36,075 = 0.3168 x 1,803,750, where 2 x 15.84 = 31.68. When the number 571428 is rotated on a circle, it stops at 288°, where 2 x 144 = 288.

$\frac{5}{7}$ = 714285 = 1485 x 481 = 297 x 2405, were 2 x 1485 = 2970.

When the number 714285 is rotated on a circle, it stops at 225°. The number 1485 is just one of twenty-four (24) permutations of 5184.

$\frac{6}{7}$ = 857142 = 1.584 x 541,125 = 0.3168 x 2,705,625, where 2 x 1.584 = 0.3168. When the number 857142 is rotated on a circle, it stops at 342°.

Please notice above that when 142857, 285714, 428571, 571428, 714285, and 85742 are rotated on a circle, the degree difference on the "stops" is always 63°. For example,

142857 produces   117° + 180° = 297°   297° - 234° = 63°.

285714 produces   54° + 180° = 234°   234° - 171° = 63°.

428571 produces   171° + 180° = 351°   171° - 108° = 63°.

571428 produces   108° + 180° = 288°   108° - 45° = 63°.

714285 produces   45° + 180° = 225°   225° - 162° = 63°.

857142 produces   162° + 180° = 342°   180° - 117° = 63°.

Please note that 36 x 144 = 5184 and 36 x 288 = 10368, where number 36 is just one (1) of two permutations of 63.

Notice that the "Speed of Light" in a vacuum = 299,792,458 meters/second, where $\frac{299,792,458}{7}$ = 42,827,494. It is highly unusual for the number 7 to divide into another number without producing the above recurring digits of 142857 (and its subsequent permutations).

It appears that the number (7) has a direct relationship with the "Speed of Light" and the "light waves" could well have a "wave pattern" similar to the "recurring digits" of the number 7. Also, the "wave angles" may be directly related to 63°. When we multiply the digits of 299,792,458, we get a result of 3,265,920 = 630 x 5184 = 315 x 10368, where 630 = 10 x 63 and 315 is one permutation of 351, which is one of the above referenced stopping points of 351°.

It is also interesting that the natural numbers 1, 2, 3, 4, 5, 6, 7, 8, 9 are directly related to our unique numbers 5184 and 10368. Multiply the digits of 1 x 2 x 3 x 4 x 5 x 6 x 7 x 8 x 9 = 362,880 = 70 x 5184 = 35 x 10368.
The last word in the Bible Holy Scriptures is "Amen", which has a Greek numeric value = 99. Notice that 16 x 99 = 1584, where 1584 is just one of 24 permutations of 5184.

Please notice too that the number (7) is directly related to 5184 and 3168 by one of their permutations. For example, 7 x 1188 = 8316 and 7 x 594 = 4158, where 8316 is a permutation of 3168 and 4158 is a permutation of 5184.

"I AM THAT I AM"

Exodus 3:14 "And God said unto to Moses, I AM THAT I AM: and he said, Thus shalt thou say unto the children of Israel, I AM hath sent me unto you".

I AM THAT I AM comes from the Hebrew words "hayah", "hayah" (Strong's word reference number 1961) and is spelled with the Hebrew letters: "Hey Yod Hey", "Hey Yod Hey", which have an English numeric value of 54 + 6 + 78 + 120 + 48 + 6 + 120 + 54 + 6 + 78 = 330 by addition and 34,681,651,200 by multiplying its digits. Notice that the Hebrew numerics value of "Hey Yod Hey", "Hey Yod Hey" equals 400 by multiplication.

The Trinity Function (34,681,651,200) = 153, which is the "image" value for "Jesus" in Greek numerics. Taking the product of these digits (34,681,651,200) we get a value of 34560. Then 3456 – 3141 = 315, which is just one of 6 permutations of 153. Note that 3141 are the first four (4) digits of PI. It is interesting that this Bible verse is found in Exodus 3:14, where the chapter and verse numbers are the first three (3) digits of Pi.

The six (6) permutations of 330 = 1332 and the six (6) permutations of 400 = 888, where 888 is the Greek numeric value for "Jesus". The Trinity Function (400) goes into an "endless, eternal loop"; it goes to 250, then to 133 and then to 55 and then it repeats again endlessly. This process would seem to indicate the "eternal and endlessness" of "Almighty God", whose name is "Jesus Christ". The Trinity Function (330) = 153 as shown above.

When we multiply 1332 x 888, we get the value of 1,182,816, where $\frac{1,182,816}{999}$ = 2368, which is the Greek numeric value for "Jesus Christ". When we add 1,182,816 + 6,182,811 = 7,365,627 = 2727 x 2701, where 2701 is the Hebrew numeric value of the entire verse in Genesis 1:1, where it says, "In the beginning, God created the heavens and the earth". This connection shows that "Jesus Christ" = "I AM THAT I AM" = "In beginning, God created the heavens and the earth", where HE is the CREATOR. Also, 7 x 3 x 6 x 5 x 6 x 2 x 7 = 52,920 = 49 x 1080, where 1080 is the Greek numeric value for "the Holy Spirit". We know that "The Spirit of God moved upon the face of the waters" (Genesis 1:2) during creation. If we calculate the twenty-four (24) permutations of 2727, we get a total value = 119,988. Then when we multiply these digits: 1 x 1 x 9 x 9 x 8 x 8 = 5184, which has direct connection to the "Total Atomic Weights of the 92 Natural Elements".

To show further connection with the total atomic weights, we use Heron's Formula for a scalene triangle, as shown below.

The area of a scalene triangle (which has three unequal sides) is found with Heron's Formula:

$A = \sqrt{S(S-x)(S-y)(S-z)}$, where x and y and z are the 3 unequal sides and S is the semi-perimeter of the triangle. Let x = 250 and y = 133 and z = 55 and $S = \frac{1}{2}(250 + 133 + 55) = 219$.

We take the values of 250, 133 and 55 from the "endless, eternal loop" mentioned above.

Then $A = \sqrt{219(219-250)(219-133)(219-55)} =$ 9785.2979515189. When we multiply these digits, we get: 9 x 7 x 8 x 5 x 2 x 9 x 7 x 9 x 5 x 1 x 5 x 1 x 8 x 9 = 5,143,824,000 = 496,125 x 10368 = 99,225 x 51840, where 51840 is directly connected to the "Total Atomic Weights of the 92 Natural Elements" and 10368 is just one permutation of 03168 = "Lord Jesus Christ". Thus, we have shown at least two (2) different connections between "I AM THAT I AM" and the creation of the natural elements. This also shows that there must by a "triangular aspect" to the created elements!

The Trinity Function (97852979515189) = 217, where 217 + 153 = 370 represents the "Power of God".

NASA Earth Facts and Figures

Please note that all the facts and figures used here were taken from: National Aeronautics and Space Administration,
http://solarsystem.nasa.gov/planets/profile.cfm?Object=Earth&Display=Facts

Earth's sidereal rotation period (length of day) = 0.99726968 earth days. Multiplying these digits: 9 x 9 x 7 x 2 x 6 x 9 x 6 x 8 = 2,939,328 = 567 x 5184 and 2 x 5184 = 10,368.

Earth's orbital circumference = 924,375,700 km. If we multiply 9 x 2 x 4 x 3 x 7 x 5 x 7, we get 52,920. Then $\frac{924,375,700}{52920} =$ 17,467.416855631. Then we multiply these digits: 1 x 7 x 4 x 6 x 7 x 4 x 1 x 6 x 8 x 5 x 5 x 6 x 3 x 1 = 101,606,400 = 19,600 x 5184 = 9,800 x 10368.

Earth's escape velocity = 40,248 $\frac{km}{hr}$ = 26 x 1548, where 1548 is just one of twenty-four (24) permutations of 5184!

And 40,248 $\frac{km}{hr}$ = 25,009 mph. Note that $(40,248)^2$ = 1,619,901,504 = 312,481 x 5184 = 1,562,405 x 1036.8.

Earth's surface area = 510,065,700 km^2. And $(5,100,657)^2$ = 26,016,701,831,649. If we multiply these individual digits, we get a value of 2,612,736 = 504 x 5184.

Earth's average distance to the sun = 149,597,890 km. When we multiply these digits (not including the zero), we get a value equal to 816,480 = 16 x 51,030, which is just one of 720 permutations of 00,153, where 153 is the "image" of Jesus in Greek numerics. Note that 8,164,800 = 1575 x 5184 = 7,875 x 1036.8.

Earth's mean orbit velocity = 107,229 $\frac{km}{hr}$. Then $107,229^2$ = 11,498,058,441. When we multiply these digits, we get 1 x 1 x 4 x 9 x 8 x 5 x 8 x 4 x 4 x 1 = 184,320. Then $184,320^2$ = 33,973,862,400 = 10,368 x 3,276,800 = 51,840 x 655,360, where 10,368 is just one of 720 permutations of 03,168 = "Lord Jesus Christ" in Greek numerics and 51,840 is just one of 720 permutations of 05,184 which is related to the "Atomic Weights of the 92 Natural Elements".

Earth's mass = 5.9737 x 10^{24}. Then $59,737^2$ = 3,568,509,169. By multiplying these digits, we get: 1,749,600 = 10.368 x 168,750 = 51.84 x 22,750 = 51,840 x 3375 = 16,875 x 10368 = 174,960,000

The Number of Days in a Solar Year

The number of days in a solar year = 365.2421896698

The Trinity Function (3652421896698) = 153, which is the "image" of "Jesus" in Greek numerics.

Now multiply the individual digits of 3652421896698 :

3 x 6 x 5 x 2 x 4 x 2 x 1 x 8 x 9 x 6 x 6 x 9 x 8 = 268,738,560 and

$\frac{268,738,560}{5184}$ = 51840, where $\frac{51840}{5}$ = 10368, which is just one of 120 permutations of 03168, which is the "Lord Jesus Christ" in Greek numerics.

Now divide 699840 = "Total Atomic Weights of all Natural Elements" multiplied digits, into 268,738,560: $\frac{268,738,560}{699840}$ = 384 and $\frac{699840}{5184}$ = 135, where 135 x 384 = 51840.

Thus, we see a direct mathematical connection between "the number of days in a solar year" and the "Total Atomic Weights of all Natural Elements" and the "Lord Jesus Christ".

Enoch's Departure

Also, "By faith Enoch was translated that he should not see death; and was not found, because God had translated him; for before his translation he had this testimony, that he pleased God". Hebrews 11:5. Notice that "translated" is the Greek word "metatithemi", which has this numerics value: Mu (30) + Epsilon (5) + Tau (300) + Alpha (1) + Tau (300) + Iota (10) + Theta (9) + Eta (8) + Mu (40) + Iota (10) = 713 by addition and 38,880 by multiplication, where 75 x 518.4 = 38,880 = 375 x 103.68.

The Elements of Life

There are Twelve Elements of Life, as listed below:

Element	Atomic Number	Atomic Weights
Hydrogen	1	1.007947
Carbon	6	12.01115
Nitrogen	7	14.0067
Oxygen	8	15.9994
Sodium	11	22.9898
Magnesium	12	24.305
Phosphorus	15	30.9738
Sulfur	16	32.064
Chlorine	17	35.453
Potassium	19	39.0983
Calcium	20	40.08
Iron	26	55.847
TOTALS	**158**	**323.83612**

When we multiply the digits of the "Total Atomic Weights" in the above example, we get:

3 x 2 x 3 x 8 x 3 x 6 x 1 x 2 = 5184, which relates directly to the "Total Atomic Weights of the 92 Natural Elements" and 2 x 5184 = 10,368, where 03168 = 3168 is just one of the 120 permutations of 10,368 and 3168 = "Lord Jesus Christ" in Greek numerics.

Also, $(158)^3$ = 3,944,312 where multiplying the digits: 3 x 9 x 4 x 4 x 3 x 1 x 2 = 2592. Note that 2592 x 2 = 5184 and 4 x 2592 = 10368.

The Hematin Molecule is composed of five (5) elements, as listed below:

Element	(Number of Atoms)	x (Atomic Weights)	=Total Atomic Weights
Iron	1	55.8452	55.8452
Nitrogen	4	14.00672	56.02688
Carbon	55	12.01078	660.5929
Hydrogen	72	1.007947	72.572184
Oxygen	5	15.99943	79.99715
TOTALS	137	98.870077	925.034314

Taking the "Atomic Weights" total of $\frac{(98,870,077)^2}{137}$ = 71,352,497,269,970. Then we multiply these digits: 7 x 1 x 3 x 5 x 2 x 4 x 9 x 7 x 2 x 6 x 9 x 9 x 7 = 360,067,680. Then 36,006,768,000 = 10,368 x 3,472,875 = 51,840 x 694,575. Also, if we take the "Total Atomic Weights" of $\frac{(92,503,431)^2}{137}$ = 62,459,012,750,159.

Then we multiply these digits:

6 x 2 x 4 x 5 x 9 x 1 x 2 x 7 x 5 x 1 x 5 x 9 = 6,804,000,

Then 680,400,000 = 10,368 x 65,625 = 51,480 x 13,125.

The Chlorophyll Molecule is composed of five (5) elements, as listed below:

Element	(Number of Atoms)	x (Atomic Weights) =	Total Atomic Weights
Magnesium	1	24.30506	24.30506
Nitrogen	4	14.00672	57.22024
Carbon	55	12.01078	660.5929
Hydrogen	72	1.007947	72.572184
Oxygen	5	15.99943	79.99715
TOTALS	137	67.329937	894.68753

Taking the "Atomic Weights" total of $\frac{(67329937)^2}{137} =$ 33,089,930,046,890. Then we multiply these digits:

3 x 3 x 8 x 9 x 9 x 3 x 4 x 6 x 8 x 9 = 33,233,088 = 5184 x 5832 = 10,368 x 2916.

Taking the "Total Atomic Weights" of $\frac{(89,468,753)^2}{137} =$ 80,046,577,633,750. Then we multiply these digits:

8 x 4 x 6 x 5 x 7 x 7 x 6 x 3 x 3 x 7 x 5 = 88,905,600 = 51840 x 1,715 = 10,368 x 8,575.

The Deoxyribonucleic Acid (DNA) relationship with the "Atomic Weights of all 92 Natural Elements"

If we use the English alphabet numeric values to replace the "instruction letters" of DNA, we get a list of "21 sets of instructions which signal 22 different functions".

DNA Instructions Chart

Name	Instructions		
Stop	UUA (258)	UAG (174)	UGA (174) = 606
Valine	GUA (174)	GUG (210)	GUC (186)
	GUU (294)	= 864	
Tyrosine	UAC (150)	UAU (258)	= 408
Tryptophan	UGG (210)	= 210	
Threonine	ACA (30)	ACG (66)	ACC 42)
	ACU (150)	= 288	
Serine	AGC (66)	AGU (174)	UCA (150)
	UCG (186)	UCC (162)	UCU (270) = 1008
Proline	CCA (42)	CCG (78)	CCC (54)
	CCU (162)	= 336	

Phenylalanine UUC (270)   UUU (378)   = 648

DNA Instructions Chart

Name	Instructions		(continued)
Lysine	AAA (18)	AAG (54)	= 72
Leucine	UUA (258)	UUG (294)	CUA (150)
	CUG (186)	CUC (162)	CUU (270) = 1320
Isoleucine	AUA (138)	AUC (150)	AUU (258) = 546
Histidine	CAC (42)	CAU (150)	= 192
Glycine	GGA (90)	GGG (126)	GGC (102)
	GGU (210)	= 528	
Glutamine	CAA (30)	CAG (66)	= 96
Glutamate	GAA (54)	GAG (90)	= 144
Cysteine	UGC (186)	UGU (294)	= 480
Asparagine	AAC (30)	AAU (138)	= 168
Aspartate	GAC (54)	GAU (174)	= 240
Arginine	AGA (54)	AGG (90)	CGA (66)
	CGG (102)	CGC (78)	CGU (186) = 576
Alanine	GCA (66)	GCG (102)	GCC (78)
	GCU (186)	= 432	

Methionine	AUG (174)	= 174

The Total Value for the DNA instruction list = 9,336 and the Total permutations of 9,336 = 139,986. Note that 27 x 5184 = 10.368 x 13,500 = 139,968, which is one of seven hundred and twenty permutations of 139,986. Also, $\frac{3168}{2}$ = 1584, which is just one of twenty-four permutations of 5184, which relates directly to the "Atomic Weight of all 92 Natural Elements". Note too that 139,968 – 9,900 = 130068, which is just one of seven hundred and twenty permutations of 003168, where 3168 = "Lord Jesus Christ".

Music Vibrations are related to the "Total Atomic Weight of the Natural Elements"

Below is a chart of the music vibrations per second, beginning with C below Middle C and progressing upward seven octaves. The numbers are on what is considered the "Just" tuning.

Vibrations per Second

C	D	E	F	G	A	B
132	148.5	165	176	198	220	247.5
264	297	330	352	396	440	495
528	594	660	704	792	880	990
1056	1188	1320	1408	1584	1760	1980
2112	2376	2640	2816	3168	3520	3960
4224	4752	5280	5632	6336	7040	7920

8448 9504 10560 11264 12672 14080 15840

The Total Vibrations (in the table above) = 180213 per second. Notice that two (2) vibrations listed above (148.5 and 1584) are both permutations of 5184 and 1584 x 2 = 3168, another vibration listed above. Also, 6336 = 2 x 3168 and 9504 = 3 x 3168 and 5 x 3168 = 15840, where 15840 is just one of 120 permutations of 51840. Notice too that all of the vibrations frequencies under D and G and B are functions of 3168 and 5184! The Trinity Function (180213) = 153, which is the "image" of "Jesus" in Greek numerics. Also, notice all of the vibration values listed under the G key: "Amen" has a Greek numeric value = 99 and all of the vibration values listed here under the Key of G are multiplies of 99!

180,213 + 312081 = 492,294, where 4 x 9 x 2 x 2 x 9 x 4 = 5184. Then 51840 divided by 5 = 10368, where 10368 is one of 120 permutations of 03168, the "Lord Jesus Christ" in Greek numerics. Note too that 3168 is one of the vibration rates in the G Note column listed above. Also, 1584 is one of 24 permutations of 5184 and 1584 is one of the vibration rates in the G Note column listed above.

Also, (180,213) x (312,081) = 56,241,053,253 and by multiplying the digits, we get 5 x 6 x 2 x 4 x 1 x 5 x 3 x 2 x 5 x 3 = 1080,000 where 1080 = "The Holy Spirit" in Greek numerics and 1080 = "God Himself who formed the earth and made it" in Hebrew numerics (see Isaiah 45:18)

Please note that the number 5184 shows up in the calculations of the "Total Atomic Weights of the 92 Natural Elements" and in the "Number of Days in a Solar Year" and in "That Certain" (see Daniel 8:13).

Atomic Time with the Cesium Atom Oscillations

The Cesium atom has 9,192,631,770 oscillations/second and is used by our government to provide accurate time.

If we multiply the above digits, we get a value of 142,884, where $\frac{142,884}{275,625} = .5184$ and $\frac{142,884}{1,378,125} = .10368$.

Also, $(142,884)^2 = 20,415,837,456$, where $\frac{20,415,837,456}{393,824,025} = 51.840$ and $\frac{20,415,837,456}{1,969,120,125} = 10.368$.

How many oscillations of the Cesium atom in one (1) year?

$9,192,631,770 \; \frac{oscillations}{second}$ times $60 \; \frac{seconds}{minute} = 551,557,906,200 \; \frac{oscillations}{minute}$

$551,557,906,200 \; \frac{oscillations}{minute}$ times $60 \; \frac{minutes}{hour} = 33,093,474,372,000 \; \frac{oscillations}{hour}$

$33,093,474,372,000 \; \frac{oscillations}{hour}$ times $24 \; \frac{hour}{day} = 7.94243384928 \times 10^{14} \; \frac{oscillations}{day}$

$7.94243384928 \times 10^{14} \; \frac{oscillations}{day}$ times $365.2421896698 \; \frac{days}{solar \; year}$
$= 2.9009119304186 \times 10^{17} \; \frac{oscillations}{solar \; year}$

If we multiply the digits (2 x 9 x 9 x 1 x 1 x 9 x 3 x 4 x 1 x 8 x 6) we get a value of 839,808, where

$$\frac{839{,}808}{162} = 5184 \text{ and } \frac{839{,}808}{81} = 10368,$$

where 5184 relates directly to the "Total Atomic Weights of the 92 Natural Elements" and 10368 is just one of 120 permutations of 03168 = "Lord Jesus Christ".

Also, if we add 9,192,631,770 + 0,771,362,919 = 9,963,994,689. Then multiplying the digits of (9,963,994,689), we get 204,073,344, where $\frac{204{,}073{,}344}{39{,}366} = 5184$ and $\frac{204{,}073{,}344}{19{,}683} = 10368$.

Heart Beat Cycles

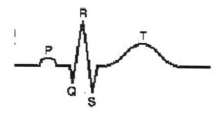

Important feature of the ECG signal (Lynch, 1985)

http://www.hcs.harvard.edu/~weber/HomePage/Papers/ECGCompression/

When medical professionals review electrocardiographs, they review the: PR Interval and the QRS Complex and the ST Segment and the QT Interval. Listed below are the wave cycles that are deemed to be "normal" within a range of times intervals.

Wave Pattern	Low Time Interval	High Time Interval	Average Time Interval
PR Interval	120 ms	200 ms	160 ms
QRS Comple	60 ms	100 ms	80 ms
ST Segment	80 ms	120 ms	100 ms
QT Interval	300 ms	440 ms	370 ms

If we take the Low Time Intervals and multiply them, we get: 120 x 60 x 80 x 300 = 172,800,000, where $(17280)^2$ = 29,859,840 = 576 x 51840 = 2880 x 10368. Also, 1728 x 3 = 5184. If we take the High Time Intervals and multiply them, we get: 200 x 100 x 120 x 440 = 1,056,000,000, where 3 x 1056 = 3168 and 15 x 1056 = 15840, which is just one of 120 permutations of 51840. If we take the Average Time Intervals and multiply them, we get: 160 x 80 x 100 x 370 = 473,600,000, where 473600000 + 000006374 = 473606374.

Then multiply the digits of 473606374 and we get a value of 254,016. Then 49 x 5184 = 254,016 and 245 x 10368 = 2,540,160. Also, $(4736)^2$ = 22,429,696. If we multiply the digits of 22,429,696, we get: 93,312 = 18 x 5184 = 9 x 10368.

So we can see from the above calculations, that "wave time intervals" (low and high and average) all produce those same numbers 51840 and 10368 as seen in other aspects of creation.

Random Analysis of Chemical Elements

In this section we want to randomly choose several chemical elements to see if the same numerical pattern exists among them. Please note that I will not be analyzing all 92 natural elements; that would greatly expand the size of this book; other researchers can have the challenge.

If we take Heron's Formula to analyze these chemical elements, we get the following: A scalene triangle has three (3) unequal sides. To find the area of this triangle, we use Heron's Formula:

$$A = \sqrt{S(S-x)(S-y)(S-z)}, \text{ where } S = \frac{1}{2}(x + y + z)$$

Let x and y and z be the three (3) unequal sides and let S equal the semi-perimeter of the triangle and let A equal the area of the scalene triangle.

A good question might be: Is there a triangular relationship among the chemical elements? Let's see if this may be the case.

Example #1 Let x = Calcium = 40.0784 atomic weight and y = Magnesium = 24.30506 atomic weight and z = Potassium = 39.09831 atomic weight. Then $S = \frac{1}{2}(400784 + 2430506 + 3909831) = 3,370,560.5$. Then (S - 400784) = 2,969,776.5 and (S – 2430506) = 940,054.5 and (S – 3909831) = - 539,270.5.

Then, $A = \sqrt{5.0744104115698 \times 10^{24}}$ = 2,252,645,203,215.5 and when we multiply these digits we get a value of 1,440,000 where $(144)^2 = 20,736 = 4 \times 5184 = 2 \times 10368$.

Notice that if we multiply the atomic weights of these 3 randomly selected chemical elements, we get:

40.0784 x 24.30506 x 39.09831 = 38,085.973300747 and when these digits are multiplied together we get a value of 106,686,720 = 2058 x 51840 = 10,290 x 10368.

Example #2 Let x = Iron = 55.8452 atomic weight and y = Rubidium = 85.46783 atomic weight and z = Silver = 107.86822 atomic weight. Then $S = \frac{1}{2}(558452 + 8546783 + 10786822)$ = 9,946,028.5. Then (S – 558452) = 9,387,576.5 and (S – 8546783) = 1,399,245.5 and (S – 10786822) = - 840,793.5.

Then $A = \sqrt{1.0984655798289 \times 10^{26}}$ = 10,480,770,867,779 and when we multiply these digits we get the value of 232,339,968 and when we multiply these digits we get a value of 419,904, where 419,904 = 81 x 5184 = 40,500 x 10368.

Notice that if we multiply the atomic weights of these 3 randomly selected chemical elements, we get:

55.8452 x 85.46783 x 107.86822 = 514,851.56873999 and when these digits are multiplied together we get a value of 2,939,328,000 = 56,700 x 51840 = 283,500 x 10368.

So after analyzing the relationships among 6 randomly selected chemical elements, it appears that there may be a triangular relationship among all of the 92 natural elements and the values of 51840 and 10368 are also prominent in these calculations.

The Ammonia molecule NH_3 was used in 1948 to create the first "atomic clock" because ammonia's single nitrogen atom flips back and forth through it triangle of 3 hydrogens at an oscillation rate of 23,870 megacycles per second. When we take $(23,870)^2 = 569,776,900$ and multiply these digits, we get a value of 714,420 = .5184 x 1,378,125 = .10368 x 6,890,625.

Photons

Physicists know that the photon only exists while moving at the "velocity of light" = 186,282 miles/second. If we take $(186,282)^2 = 34,700,983,524$. Then $\frac{34,700,983,524}{66,938,625,625} = 0.51840$ and $\frac{34,700,983,524}{334,693,128,125} = 0.10368$.

Another way of looking at this is to take the "velocity of light in a vacuum" = 299,792,458 meters/second. Then $(299,792,458)^2 = 8.9875517873682 \times 10^{16}$. Then $\frac{8.9875517873682 \times 10^{16}}{17,337,098,355,263} = 5184$ and $\frac{8.9875517873682 \times 10^{16}}{86,685,491,776,315} = 1036.8$ Thus, we see that the photon's existence is directly related to the numbers 5184 and 10368.

"That Certain..." Daniel 8: 13   The Wonderful Numberer

Dan 8:13 Then I heard ^{08085} one ^{0259} saint ^{06918} speaking ^{01696}, and another ^{0259} saint ^{06918} said ^{0559} unto that certain ^{06422} [saint] which spake ^{01696}, How long [shall be] the vision ^{02377} [concerning] the daily ^{08548} [sacrifice], and the transgression ^{06588} of desolation ^{08074}, to give ^{05414} both the sanctuary ^{06944} and the host ^{06635} to be trodden under foot ^{04823}? פלמוני palmowniy {pal-mo-nee'} probably for 06423; TWOT - 1772a; pronoun AV - that certain 1;1 1) a certain one. Numeric value: 216.

(See the Blue Letter Bible).   English transliteration: "La Palmoni" or "Wonderful Numberer".

Numeric value by addition   Pe(80) + Lamed (30) + Mem (40) + Vau (6) + Num (50) + Yod (10) = 216 = 6 x 6 x 6 (Hebrew letter values for "that certain").   This is the value of "Palmoni" without the prefix "La".

The Trinity Function (216) = 153, which is the "image of Jesus" in Greek numerics and the six (6) permutations of 216 = 216 + 261 + 126 + 162 + 621 + 612 = 1998, where 1 x 9 x 9 x 8 = 648. 16 x 648 = 10368 and 80 x 648 = 51840.

Numeric value by multiplication   Pe (80) x Lamed (30) x Mem (40) x Vau (6) x Num (50) x Yod (10) = 288,000,000 (Hebrew letter values for "that certain"). Note: 3168 (Lord Jesus Christ in Greek numerics) minus 2880 = 288 = 2 x 144.

This is the value of "Palmoni" without the prefix "La". The Total Value for the 24 permutations of 2880 = 119988 and 1 x 1 x 9 x 9 x 8 x 8 = 5184 = 8 x 648.

The Total Value for the 24 permutations of 3168 = 119988 and 1 x 1 x 9 x 9 x 8 x 8 = 5184 = 8 x 648. Note that 2 x 5184 = 10368. Note that the Total Atomic Weights of all 92 natural elements = 10,310.6645966, where the product of these digits = 1 x 3 x 1 x 6 x 6 x 4 x 5 x 9 x 6 x 6 = 135 x 5184.

Notice that 135 is one of six (6) permutations of 153 the Greek numeric value for "image value of Jesus".

Thus, the Total Value for the 24 permutations of 2880 (which results from multiplying the Hebrew letter values of "that certain") and the Total Value for the 24 permutations of 3168 and the Total Atomic Weights of all 92 natural elements are all the mathematically related to the number 5184! So we can conclude that "that certain" person is the Lord Jesus Christ who created the elements!

Please note that the Total Permutations Value for 5184 and 2880 and 3168 all equal 119988. Thus, the "Total Atomic Weights of all 92 natural elements" has a mathematical factor (i.e. 5184) that has the same permutations value as for 3168 and 2880.

Notice too that 5184 is just one of 24 permutations of $1584 = \frac{3168}{2}$.

"La Palmoni" means "the secret numberer". The Hebrew numeric value for "La Palmoni" is: Lamed (30) + Pe(80) + Lamed (30) + Mem (40) + Vau (6) + Num (50) + Yod (10) = 246 by addition and = 8,640,000,000 by multiplication. When we take $(8640)^2$ = 74,649,000 = 1440 x 51840 = 7200 x 10368.

Isaiah 9:6 says that "-----his name shall be called Wonderful---------"The word "wonderful" is Strong's number 6382, which is just one of 24 permutations of 2368 the Greek numeric value for "Jesus Christ".

The Hebrew word for "wonderful" is "pe-le" or "peh-leh" and has a numeric value of aleph (1) + Lamed (30) + Pe (80) = 111 by addition. Notice that 111 represents the Trinity of 3 equal persons; i.e. the "Godhead".

The Trinity Function (111) = 153 the Greek numeric value for the "image of Jesus". The six (6) permutations of 111 = 666 by addition, where 666 = 18 x 37 and 6 x 6 x 6 = 216, the Hebrew numeric value (by addition) of "that certain" in Daniel 8:13.

Prime Numbers and the Atomic Weights of the 92 Natural Elements

Listed below is a Prime Number Table laid out in column and rows where the numbers are assigned to Column 1 or Column 2 or Column 3 or Column 5 or Column 7 or Column 9 according to the last digit of the prime number; this is not the usual display of the prime numbers.

Prime Number Counting Grid

Row #	Column 1	Column 2	Column 3	Column 5	Column 7	Column 9
1	1	2	3	5	7	9
2	11		13		17	19
3	31		23		37	29
4	41		43		47	59
5	61		53		67	79
6	71		73		97	109
7	101		83		107	139

The purpose of laying out the prime numbers according to this scheme is to clearly see the patterns develop. Please note that this prime number table could be carried out to infinity; obviously, I don't have room in this book to do so. Some very interesting things happen when the prime numbers are displayed in a table as illustrated above.

Look what happens: Draw a circle with six (6) radii emanating from the center point, where each radius represents one of the six (6) columns above.

Now using the Creator's "human hand geometry", spread your "thumb and index finger" as far apart as possible, and if you are "normal" like most people, you will see that your "maximum possible spread between the thumb and index finger is approximately 144 degrees on the circle" and the "maximum spread between the index finger and the middle finger is about 72 degrees" and the "maximum spread between the middle finger and the ring finger is about 36 degrees" and the "maximum spread between the ring finger and the little finger is about 36 degrees. So based on this "human hand geometry", draw the radii of the circle such that "Column 1" radius is 36 degrees from the 0 degree point, and draw "Column 2" radius is 36 degrees from "Column 1" radius and draw "Column 3" radius is 36 degrees from "Column 2" radius and draw "Column 5" radius is 72 degrees from "Column 3" radius and draw "Column 7" radius 72 degrees from "Column 5" radius and draw "Column 9" radius 72 degrees from "Column 7" radius. Notice in the table above that each row of primes is 10 or 20 or 30 higher or higher than the previous row. For example, in column 1, row 1 we have the prime number "1" and in column 1, row 2 we have the prime number 11 and in column 1, row 3 we have the prime number 31 and continuing this way in all of the other columns and rows.

Now draw a series of concentric circles outward from your original circle (similar to the waves you see when dropping a stone in the water) and keeping your original radii in place; you will make each circle such that "prime number 1" is plotted on "Column 1" on the original circle, and "prime number 11" is plotted on "Column 1" on the first concentric circle, and "prime number 31" is plotted on "Column 1" on the second concentric circle and etc out to infinity for all of the prime numbers.

Now plot all of the prime numbers in each column unto their respective radii and concentric circles; you can do this for all primes out to infinity.

Now as you observe your drawing, you can see that you only have one (1) entry on "Column 2" radius for the "prime number 2" and only one (1) entry on "Column 5" radius for the "prime number 5". From here on out to infinity, there are no other prime numbers on the "Column 2" radius or the "Column 5" radius, because no other prime numbers end with the digit "2" or "5". So only "Column 1" and "Column 3" and "Column 7" and "Column 9" radii will plot prime numbers out to infinity and these four (4) columns form a "CROSS" where the radii extend to infinity! I think this "CROSS" represents the cross of Jesus Christ because as every Christian knows, the "cross of Jesus Christ" is central to Christianity. And it has been shown repeated throughout this book, that the number 3168 = "Lord Jesus Christ" is present in all aspects of the physical and mathematical realms. The Prime Numbers are like the "backbone" of the "natural numbers".

You will also notice from your drawing above, that all of the prime numbers (as plotted on the circles above) have a "triangular relationship" among themselves.

For example, prime number 71 is plotted on the "Column 1" radius and prime number 73 is plotted on the "Column 3 radius, where the angle between these radii is 72 degrees. So we know that this "triangle" has two (2) known sides: 71 and 73 and the angle between these sides is 72 degrees. In fact, any prime numbers that have their last digit as a "1" or as a "3" would have a "72 degree relationship" between them. So, for example, prime number 501821 and prime number 918173 would have a "72 degree relationship" between them.

Using the Ultimate Triangle Calculator:

http://www.1728.com/trig4.htm , we have these given facts:

Side a = 71 and Angle C = 72 degrees and Side b = 73. Using this online calculator, we get these values: Angle A (opposite Side a) = 52.905 degrees and Side c = 84.657 and Angle B (opposite Side b) = 55.095 degrees and the Area of the triangle = 2.4647 x10^3. This procedure can be used to find an infinite number of triangular relationships among the prime numbers.

Please note that if we take $(84657)^2$ = 7,166,807,649, where these multiplied digits equal 3,048,192 = 588 x 5184 = 294 x 10368.

Please note that if we take $(52905)^2$ = 2,798,939,025, where these multiplied digits equal 2,449,440 = 518.4 x 4725 = 103.68 x 23,625.

Thus, we see that these prime numbers have a triangular relationship among themselves and these prime numbers have a direct relationship with the "Total Atomic Weights of the 92 Natural Elements" and a relationship with the "LORD JESUS CHRIST" = 3168.

It appears to me that these prime numbers (using the above format and procedures) form a "Conical Helix" and may useful in nuclear physics.

Aaron's Breastplate Exodus 28: 15 – 21

Shown below is a 4 x 3 matrix of the Breastplate giving the sons of Jacob and their numerical value of their Hebrew names, which are listed from right to left, in the Hebrew manner – with the omitted names of Levi and Joseph.

Judah (30)	Simeon (466)	Reuben (259)
Gad (7)	Naphtali (570)	Dan (54)
Zebulun (95)	Issachar (830)	Asher (501)
Ephraim (331)	Manasseh (395)	Benjamin (162)

If we lay out the numerical values of these sons name in a straight row going from right to left as shown above, we get:

2594663054570750183095162395331

Now we list the digits of Pi: 31415926535897932384626433832 79

By matching up the digits of these numbers, we get a value for Pi = 3.14159265358979323 with a 17 digit accuracy! The left over digits from the sons' names are: 04075005531. If we multiply the above digits of Pi, we get the value of 39,680,298,000 = 765,450 x 51840 = 3,827,250 x 10368. If we multiply the digits of the sons' numerical values, we get a value of $1.85177664 \times 10^{14}$ = 3,572,100,000 x 51840 = 17,860,500,000 x 10368.

If we add the above values of the sons' names and the value for Pi, we get the value of $2.5978046472605 \times 10^{30}$ and when we multiply these digits, we get 203,212,800 = 3920 x 51840 = 19,600 x 10368.

It seems that no matter how these numbers are analyzed, they yield the values of 51840 and 10368 which have been shown to have many mathematical and physical relationships through out this book.

The 12 foundations of the Heavenly City of New Jerusalem

> *And the foundations of the wall of the city [were] garnished with all manner of precious stones. The first foundation [was] jasper; the second, sapphire; the third, a chalcedony; the fourth, an emerald; the fifth, sardonyx; the sixth, sardius; the seventh, chrysolite; the eighth, beryl; the ninth, a topaz; the tenth, a chrysoprasus; the eleventh, a jacinth; the twelfth, an amethyst. (Revelation 21:19-20)*

- Jasper   Strong's Number 2393   Numeric Value = 120
- Sapphire   Strong's Number 4552   Numeric Value = 174
- Chalcedony   Strong's Number 5472   Numeric Value = 69
- Emerald   Strong's Number 4665   Numeric Value = 820
- Sardonyx   Strong's Number 4557   Numeric Value = 380
- Sardius   Strong's Number 4556   Numeric Value = 588
- Chrysolyte   Strong's Number 5555   Numeric Value = 749
- Beryl   Strong's Number 969   Numeric Value = 66
- Topaz   Strong's Number 5116   Numeric Value = 61
- Chrysoprosus   Strong's Number 5556   Numeric Value = 200
- Jacinth   Strong's Number 5192   Numeric Value = 89
- Amethyst   Strong's Number 271   Numeric Value = 16

Total Numeric Value for All 12 Stones = 3332, where $(3332)^3$ = 36,992,6<u>10,368</u>; then multiplying these digits, we get a value = 2,519,424 = 486 x 5184 = 243 x 10368.

The above Hebrew and Greek definitions were taken from a Bible dictionary and I calculated the numeric values. Strong's Exhaustive Bible Concordance.

Truth: What Is It?

- Merriam-Webster's Collegiate Dictionary (Tenth Edition) says that Truth is: "Fidelity, constancy; sincerity in action, character, and utterance; the body of real things, events and facts; a transcendent fundamental or spiritual reality".
- Theophysic Gematria written by Dennis Lee Oberholtzer (see the bibliography) says that the Holy Bible Scriptures use four (4) different words for Truth. I will be evaluating these two (2) different Hebrew words and two (2) different Greek words in this section of the book, many thanks to my good Cousin Dennis!

Please note that the Hebrew and Greek words will be evaluated here by using a free online Holy Bible database.

1) 'emeth {eh'-meth} contracted from 0539; TWOT - 116k AV - truth 92, true 18, truly 7, right 3, faithfully 2, assured 1, assuredly 1, establishment 1, faithful 1, sure 1, verity 1; 127 n f 1) firmness, faithfulness, truth 1a) sureness, reliability 1b) stability, continuance 1c) faithfulness, reliableness 1d) truth 1d1) as spoken 1d2) of testimony and judgment 1d3) of divine instruction 1d4) truth as a body of ethical or religious knowledge 1d5) true doctrine adv 2) in truth, truly.

Genesis 42:16 Send one of you, and let him fetch your brother, and ye shall be kept in prison, that your words may be proved, whether *there be any* truth in you: or else by the life of Pharaoh surely ye *are* spies.

1 Samuel 12:24 Only fear the LORD, and serve him in truth with all your heart: for consider how great *things* he hath done for you.

Notice that the Hebrew numeric value for truth = 441 and this word appears 23 times in the Old Testament scriptures. Therefore, 441 x 23 = 10143 and $(10143)^2$ = 102,880,449. If we add 102,880,449 + 944,088,201 = 1,046,968,650, then multiply these digits, we get a value of 311,040 = 60 x 5184 = 30 x 10368. Note that 10368 is just one of a 120 permutations of 03168 = "Lord Jesus Christ" in Greek numerics. Also notice that 441 has the "reverse digits" of 144, where 36 x 144 = 5184 and 72 x 144 = 10368.

2) 'amen {aw-mane'} from 0539; TWOT - 116b; adverb AV - amen 27; truly 2, so be it 1; 30 1) verily, truly, amen, so be it.

Deuteronomy 27:15 Cursed *be* the man that maketh *any* graven or molten image, an abomination unto the LORD, the work of the hands of the craftsman, and putteth *it* in *a* secret *place*. And all the people shall answer and say, Amen.

Isaiah 65:16 That he who blesseth himself in the earth shall bless himself in the God of truth; and he that sweareth in the earth shall swear by the God of truth; because the former troubles are forgotten, and because they are hid from mine eyes.

Notice that the Hebrew numeric value of truth = 741 and this word appears 24 times in the Old Testament scriptures.

Notice that $(741)^3$ = 406,869,021 and these digits multiply to a value = 20,736 = 4 x 5184 = 2 x 10368. In Deuteronomy 27:15-26 there are 12 "Amen" s found in these verses. The number 12 is used many times through out the Holy Bible scriptures referring to "government".
Notice that 12 x 741 = 8892, where $(8892)^2$ = 79,067,664 = 51.85 x 1,525,225 = 10.368 x 7,626,125.

3) aletheia {al-ay'-thi-a} from 227; TDNT - 1:232,37; n f
AV - truth 107, truly + 1909 1, true
1, verity 1; 110 1) objectively 1a) what is true in any matter under consideration 1a1) truly, in truth, according to truth 1a2) of a truth, in reality, in fact, certainly 1b) what is true in things appertaining to God and the duties of man, moral and religious truth 1b1) in the greatest latitude 1b2) the true notions of God which are open to human reason without his supernatural intervention 1c) the truth as taught in the Christian religion, respecting God and the execution of his purposes through Christ, and respecting the duties of man, opposing alike to the superstitions of the Gentiles and the inventions of the Jews, and the corrupt opinions and precepts of false teachers even among Christians 2) subjectively 2a) truth as a personal excellence 2a1) that candor of mind which is free from affection, pretence, simulation, falsehood, deceit.

John 1:14 And the Word was made flesh, and dwelt among us, and we beheld his glory, the glory as of the only begotten of the Father, full of grace and truth.

1John 1:6 If we say that we have fellowship with him, and walk in darkness, we lie, and do not the truth.

Notice that the Greek numeric value for truth = 64 and this word appears 104 times in the New Testament scriptures. Therefore, 104 x 64 = 6656, where 6 x 6 x 5 x 6 = 1080 = "the Holy Spirit" in Greek numerics. Also, $(6656)^2$ = 44,302,336 and when you multiply these digits, you get a value equal to 5,018,004 and 2 x 5,018,004 = 10,036,008.

Notice that 5,018,004 is really one of 40,320 permutations of 0005184 and 10,036,008 is just one of 362,880 permutations of 00003168 and the 362,880 permutations have imbedded the value for 2368 = Jesus in Greek numerics.

4) aletheian {al-ay'-thi-an} from 227; TDNT - 1:232,37; n f AV - truth 107, truly + 1909 1, true 1, verity 1; 110 1) objectively 1a) what is true in any matter under consideration 1a1) truly, in truth, according to truth 1a2) of a truth, in reality, in fact, certainly 1b) what is true in things appertaining to God and the duties of man, moral and religious truth 1b1) in the greatest latitude 1b2) the true notions of God which are open to human reason without his supernatural intervention

1c) the truth as taught in the Christian religion, respecting God and the execution of his purposes through Christ, and respecting the duties of man, opposing alike to the superstitions of the Gentiles and the inventions of the Jews, and the corrupt opinions and precepts of false teachers even among Christians 2) subjectively 2a) truth as a personal excellence 2a1) that candor of mind which is free from affection, pretence, simulation, falsehood, deceit.

Notice that this fourth definition of "Truth" is very similar to aletheia {al-ay'-thi-a} above in 3); here Truth is defined as "verifiable truth". The numeric value of aletheian = 114.

2 John 1:1 The elder unto the elect lady and her children, whom I love in the truth; and not I only, but also all they that have known the truth; 2 John 1:2 For the truth's sake, which dwelleth in us, and shall be with us for ever.

Here we see in verse 2 that truth dwells in us; that is verifiable and indicates that the Truth is not only "the body of real things, events and facts" but also a Person = "Lord Jesus Christ".
As stated above, the number 12 is used for "government" through the Holy Bible scriptures.
Notice that 12 x 114 = 1368, which is just one of twenty-four permutations of 3168 = "Lord Jesus Christ". Also, since this word appears 104 times in scripture, we take 114 x 104 = 11,856 = 16 x 741, which is the numeric value of "Amen" (a Hebrew word for Truth) listed above.

If we take the numeric values of the four (4) scriptural words for "Truth", we get $(64 \times 114 \times 441 \times 741) = 2{,}384{,}194{,}176 = 5184 \times 459{,}914 = 10368 \times 229{,}957$. And if we take $(64 + 114 + 441 + 741 = 1360$, where the twenty-four (24) permutations of $1360 = 66660$.

Then $(6666)^2 = 44{,}435{,}556$ and these digits multiplied have a value $= 144{,}000$ and $(144)^2 = 20{,}736 = 4 \times 5184 = 2 \times 10368$.

Thus, we see from the above Biblical discussion of "Truth" that "spiritual truth" has direct relationship with "physical and mathematical truth" as shown throughout this book. Indeed there is only one source for "Truth" and that is found in the Person of the Lord Jesus Christ!

Hummingbirds!

The "World of Hummingbirds" is a very interesting website that lists many facts about hummingbirds; I was especially interested in the "heart beat frequencies" and the "wing beat frequencies".

A hummingbird's heart beats up to 1260 times per minute. Then $(1260)^3 = 2{,}000{,}376{,}000 = 5184 \times 385{,}875 = 1036.8 \times 1{,}929{,}375$.

A hummingbird's wings beat up to 200 times per second. Then (200 times/second) times (60 seconds) = 12,000 beats per minutes. $(12)^7 = 35{,}831{,}808 = 5184 \times 6912 = 10368 \times 3456$.

Thus, we see again that 5184 and 10368 coordinate with many other aspects of our physical and mathematical and spiritual universe!

Nuclear Explosion Facts

The following facts were taken from the: http://www.geocities.com/CapeCanaveral/Lab/2742/numbers.htm website and provide interesting confirmation with the rest of the material in this book.

Nuclear explosions typically happen in just $\frac{1}{1,000,000}$ of a second at a temperature of up to 18,000,000 degrees Fahrenheit and the resulting EMP (electromagnetic pulse), which is light, heat, sound and blast, rises at about 480 kilometers/hour = 298.258 miles/hour. Using these figures, we come up with some interesting results: $(18)^7$ = 612,220,032 = 5184 x 118,098 = 10368 x 59,049.

When we multiply the digits of 298.258, we get the result equal to 11,520. Then taking $(1152)^2$ = 1,327,104 = 5184 x 256 = 10368 x 128. It is amazing to see these same two (2) numbers continually relate to all other physical and mathematical relationships.

Pi (π) and Golden Ratio (ϕ) and Euler's e and Number 153

These three (3) mathematical constants and the Number 153 are also directly related to the numbers 5184 and 10368 as I will show below:

Pi (π) = 31,415,926,535,898 without the decimal point. Take $\frac{31,415,926,535,898}{60,601,710,138,692}$ = 0.5184 and $\frac{31,415,926,535,898}{30,300,855,069,346}$ = 1.0368. If we take the product of the digits 60,601,710,138,692, we get a value equal to 653,184 = 126 x 5184 = 63 x 10368.

And if we take the product of the digits 30,300,855,069,346, we get a value equal to 6,998,400 = 1350 x 5184 = 675 x 10368.

Golden Ratio (φ) = 161803398874989 without the decimal point. Take $\frac{161803398874989}{31,212,075,400,268}$ = 5.184 and $\frac{161803398874989}{15,606,037,700,134}$ = 10.368. If we take the product of the digits of 31,212,075,400,268, we get a value equal to 161,280 = 512 x 315 = 16 x 10080, where 315 is just one permutation of 153, which is the "image" of Jesus in Greek numerics. Note that 10080 is the approximate atomic weight of Hydrogen. If we take the product of the digits of 15,606,037,700,134, we get a value equal to 317,520 = 6125 x 51.84 = 30,625 x 10.368.

Euler's e = 2718281828459 without the decimal point. Take $\frac{2718281828459}{52,435,992,061,324}$ = 0.5184 and $\frac{2718281828459}{26,217,996,030,662}$ = 1.0368. If we take the product of the digits of 52,435,992,061,324, we get a value equal to 13,996,800 = 2700 x 5184 = 1350 x 10368. If we take the product of the digits of 26,217,996,030,662, we get a value equal to 17,635,968 = 3402 x 5184 = 1701 x 10368.

Number 153 is well known as the "image" of Jesus in Greek numerics. If we take the product of the six permutations of 153, we get: $(153 \times 135 \times 315 \times 351 \times 513 \times 531)$ = 6.2209219959023 x 10^{14}. Take $\frac{62,209,219,959,023}{30,000,588,329,004}$ = 5.184 and $\frac{62,209,219,959,023}{60,001,176,008}$ = 10.368. If we take the product of the digits of 30,000,588,329,004, we get a value equal to 207,360 = 40 x 5184 = 20 x 10368. If we take the product of the digits of 60,001,176,658,008, we get a value of 483,840 = 315 x 1536 = 48 x 10080, where 315 is just one permutation of 153 = the "image" of Jesus in Greek numerics. Note that 10080 is the approximate atomic weight of Hydrogen.

It appears to me that the above mathematical constants (even if carried out to an infinite number of digits) may have 5184 and 10368 (and any of their permutations) as even divisors.

I think it is very interesting that these important constants have similar divisors; this should have enormous implications for mathematicians and scientists. I have been using a hand-held calculator for all of the calculations done in this book. Obviously, if anyone has an access to a supercomputer, there is much more that can be done and analyzed.

Pi divided by 5184 and 10368

31415926535897932384626433832795028841971693993751 0

Divided by
5184000000000000000000000000000000000000000000000.

The result is:

6.06017101386919991987392627947434969945441627965875771 6 04938<u>271604938</u>……..

Notice that the digits <u>271604938</u> repeat continuously out to infinity!

If the above Pi digits are simply divided by 5184, you get <u>135802469</u> as the repeating digits out to infinity.

3141592653589793238462643383279502884197169399375 10

Divided by 10368 =

3.03008550693459995993696313973717484972720813982937 8858 02469135802469……….

Notice that the digits of 135802469 repeat continuously out to infinity! If the above Pi digits are simply divided by 10368, you get 123456790 as the repeating digits out to infinity!

It appears to me that the digits 5184 and 10368 control the distribution of the Pi digits and the "terminal repeating digits" vary with the number of Pi digits being divided.

Pi (π) is an irrational number, which means that it cannot be expressed as a fraction m/n, where m and n are integers. Consequently its decimal representation never ends or repeats.

Beyond being irrational, it is a transcendental number, which means that no finite sequence of algebraic operations on integers (powers, roots, sums, etc.) could ever produce it. http://en.wikipedia.org/wiki/Pi (Note this is a standard mathematical definition for an irrational number).

Do the above examples demonstrate that Pi (π) is NOT an irrational number because 1) it can be expressed as a fraction m/n and 2) its decimal representation repeats? I have divided 51 digits of Pi (π) by 5184 x 10^{+46} and got the repeating digits of 271604938 and I have divided 51 digits of Pi (π) by 10368 x10^{+46} and got the repeating digits of 135802469. Please notice that the repeating decimal in the first example above (271604938) has a "digits product" value equal to 72,576 = 14 x 5184 = 7 x 10368.

And the repeating decimal in the second example above (135802469) has a "digits product" value equal to 51,840 = 10 x 5184 = 5 x 10368.

Pi = 31415926535897932384626433832795028841971693993751058209749445923078164 (without the decimal point)

Divided by 2.701 =

1.16312204871891641557298903490540647323108826337471522435207130407545960755275<u>823768974453905960755275</u>………..

Please notice the pattern of repeating digits out to infinity: <u>823768974453905960755275</u>; there are 24 repeating digits and when we multiply these digits, we get the value equal to 7.2589644288 x 10^{15}.

Then $\frac{7.2589644288 \times 10^{15}}{1,400,263,200,000}$ = 5184 and $\frac{7.2589644288 \times 10^{15}}{700,131,600,000}$ = 10368

Please note that 2701 is a very special Biblical number, because it exactly equals the numerical value of Genesis 1:1, which says, "In the beginning God created the heaven and the earth". Thus, we see that this "God Number" = 2701 "controls the distribution of Pi digits by dividing Pi by 5184 or 10368, which is just one of 120 permutations of 03168 = "Lord Jesus Christ" in Greek numerics!! These "24 repeating digits" represent exceedingly complex permutation calculations! These "24 repeating digits" can be "permutated" 6.2044840173324 x 10^{23} times!! This truly an astronomical number!!

Listed here are some of the mathematical relationships between 5184 and 10368:

2 x 8415 = 16830 2 x 5184 = 10368 2 x 1584 = 3168 2 x 5418 = 10836 2 x 4158 = 8312 2 x 8154 = 16308. Notice that this numbers are all permutations of 5184 and 10368.

What happens if we divide Pi (reversing the digits) by 2701?

4618703295449479028501573993961791488205972383346264832 3979853562951413

Divided by 2701 =:

1.70999751775249131007092706181480617852868285203489997497148661839879352091817845242502776749352091817845242502776749352091817845242500..........

Here we see the same phenomenon as in the above illustration; when Pi or Reverse Pi digits are divided by 2701 we get 24 repeating digits (277674935209181784524250) and when we multiply these digits, we get a value equal to 1.18294235136 x 10^{14}.

Then we get; $\frac{1.18294235136 \times 10^{14}}{22,819,104,000}$ = 5184 and $\frac{1.18294235136 \times 10^{14}}{11,409,552,000}$ = 10368. Again we see that the God Numbers of 2701 and 5184 and 10368 control the distribution of Pi digits and Reverse Pi digits.

These calculations show a direct mathematical relationship between Pi and the Creator God in Genesis 1:1!

$\sqrt{2}$ =
1.41421356237309504880168872420969807856967187537694

Now divide:

14142135623730950488016887242096980785696718753 7694 by

5184000000000000000000000000000000000000000000000.

The result is:

2.72803542124439631327486250812055965142113957131 44290

<u>123456790</u>......... (These digits repeat continuously out to infinity!)

.......Notice that when I multiply the digits of <u>123456790</u> a I get the value equal to 45,360 = 8.75 x 5184.

Now take:
14142135623730950488016887242096980785696718753 7694

Divided by
10368000000000000000000000000000000000000000000 =

13.64017710622198156637431254060279782571056978 56572145061728 <u>395061728</u>…….. (repeating digits out to infinity!)

Notice that when I multiply the digits of <u>395061728</u>, I get the value equal to 90,720 = 17.5 x 5184.
Notice above that $\sqrt{2}$ (51 digits) was divided by 5184 x 10^{+46} and it has produced a rational number with repeating decimal digits 123456790, where the product of these digits = 45,360 = 0.5184 x 87,500 = 0.10368 x 437,500.

And that $\sqrt{2}$ (51 digits) was divided by 10368 x 10^{+45} and it has produced a rational number with repeating decimal digits 395061728, where the product of these digits = 90,720 = 518.4 x 175 = 103.68 x 875.

Do the above examples demonstrate that $\sqrt{2}$ is NOT an irrational number because 1) it can be expressed as a fraction m/n and 2) its decimal representation repeats?

Looking at the Structure of the Prime Numbers

In my analysis of the Prime Numbers, I have prepared six (6) examples:

Example #1 The Prime Numbers from 1 to 137 with all of the "spaces" between them eliminated and divided by 5184

Example #2 The Prime Numbers from 137 to 1 with all of the "spaces" between them eliminated and divided by 5184

Example #3 The Prime Numbers from 1 to 137 with all of the "spaces" between them eliminated and divided by 10368

Example #4 The Prime Numbers from 137 to 1 with all of the "spaces" between them eliminated and divided by 10368

Example #5 The Prime Numbers from 2 to 137 with all of the "spaces" between them eliminated and divided by 5184

Example #6 The Prime Numbers from 137 to 2 with all of the "spaces" between them eliminated and divided by 5184

Example #1

12357111317192329313741434753596167717379838997101103107109113127131137 divided by 5184

The result is:

2.38370202877938451268160392623382864918592573246549056850098632853609895833333333…..(repeating digits out to infinity!) Notice that when the repeating digits 333333333 are multiplied together, we get a value equal to 19,683 = 0.005184 x 3,796,875.

Example #2

7311317213119017013011017998389737177616953574341473139 2329171311175321 divided by 5184

The result is:

1.410362116728205442324656249689378313583517279001055775 31499173053964739583333333333……(repeating digits out to infinity!)

Notice that when the repeating digits 333333333 are multiplied together, we get a value equal to 19,683 = 0.005184 x 3,796,875.

Example #3

12357111317192329313741434753596167717379838997101103107109113127131137 divided by 10368. The result is:

1.1918510143896922563408019631169143245929628662327452842504931642680494791666666666…..(repeating digits out to infinity!) Notice that when we multiply together the repeating digits 666666666, we get a value equal to 10,077696 = 1944 x 5184.

Example #4

731131721311901701301101799838973717761695357434147313923291
71311175321 divided by 10368

The result is:

7.0518105836410272116232812484468915679175863950052788765749586526982369791<u>666666666</u>…..(repeating digits out to infinity!) Notice that when we multiply together the repeating digits <u>666666666</u>, we get a value equal to 10,077696 = 1944 x 5184.

Example #5

235711131719232931374143475359616771737983899710110310710911 3127131137 divided by 5184

The result is:

4.546896831003721670025915805548163035069133867864782228219739858570866126 54320<u>987654320</u>…….(repeating digits out to infinity!) Notice that when we multiply together the repeating digits of <u>987654320,</u> we get the value of 362,880 = 70 x 5184.

Example #6

731131721311901701301101799838973717761695357434147313923291
7131117532 divided by 5184. The result is:

1.41036211672820544232465624968937831358351727900105577531499173053964737 654320<u>987654320</u>….(repeating digits out to infinity!) Notice that when we multiply together the repeating digits of <u>987654320,</u> we get the value of 362,880 = 70 x 5184.

Many mathematicians think that the number 1 should be considered to be "a prime number"; I agree with that analysis. As you can see from the illustrations below, when number 1 is considered to be "prime", then the division by the numbers 5184 and 10368 give "cleaner" results.

In either case, these illustrations show that the number 5184 and 10368 can "control the generation" of the prime numbers. I truly think that if all prime numbers out to infinity were calculated in a similar fashion, the numbers 5184 and 10368 would still "control the generation" of the prime numbers. I leave this challenge to scientists and mathematicians who have super computer computational power to prove or disprove this theory.

Fibonacci Numbers are related to 5184 and 10368

As shown in the examples above with the Prime Numbers and Pi (π) and $\sqrt{2}$, appears that 5184 and 10368 (and their various permutations) control the generation of Fibonacci Numbers too. In the two (2) examples below, notice that I have taken the first twenty-four (24) Fibonacci Numbers and eliminated the "space between them" to create one large number and then divided 5184 and 10368 into this large number. Listed below are the results of these calculations:

112358132134558914423377610987159725844181676510946177112865746368 divided by 5184 =

2.16740224024997905909216360144128033534757267845475728609734904777777777..................(continuously repeating digits out to infinity!).

1123581321345589144233377610987159725844181676510946177
112865746368 divided by 10368 =

1.0837011201249895295460818007206401676737863392273786430486745238888888888......(continuously repeating digits out to infinity!).

Thus, we see that 5184 and 10368 seem to control the generation of the Fibonacci Numbers!

Euler's e and Feigenbaum's Constant and

The Numbers 5184 and 3168

In this section I want to show that the numbers 5184 and 3168 seem to also control the generation of Euler's e value and Mitchell Feigenbaum's Constant.

Euler's e =
2.71828182845904523536028747135266249775724709369995957496696 76277240766303535 divided by 5184 equals:

0.00052435992061324175064820360172697964848712328196372676986 24551751010950290033757 7160493827.......(continuously repeating digits out to infinity!).

Notice that when the repeating digits of 160493827 (which is a permutation of 271604938) are multiplied, we get a value equal to 72,576 = 14 x 5184.

Euler's e =
2.71828182845904523536028747135266249775724709369995957496696 76277240766303535 divided by 10368 equals:

0.00026217996030662087532410180086348982424356164098186338493 12275875505475145016878858024 69135802469..............
(Continuously repeating digits out to infinity!).

Notice that when we multiply the repeating digits of <u>135802469</u>, we get a value equal to 51,840.

Eulers's e=
2.71828182845904523536028747135266249775724709369995957496696 76277240766303535 divided by 3168 equals:

0.00085804350645803195560615134828051215206983809775882562341 12902865290645929146148<u>98989898</u>……(continuously repeating digits out to infinity!).

Notice that when we multiply the repeating digits of <u>898989898</u>, we get a value equal to 214,990,848 = 41,472 x 5184. Remember that the number 3168 equals the numeric value for "Lord Jesus Christ".

The Feigenbaum constants are two <u>mathematical constants</u> named after the mathematician <u>Mitchell Feigenbaum</u>. Both express ratios in a <u>bifurcation diagram</u>.

δ= 4.66920160910299067185320382…….

Feigenbaum originally related this number to the period-doubling bifurcations in the <u>logistic map</u>, but also showed it to hold for all one-dimensional maps displaying a single hump. As a consequence of this generality, every chaotic system that corresponds to this description will bifurcate at the same rate. Feigenbaum's constant can be used to predict when chaos will arise in such systems before it ever occurs. It was discovered in 1975.

The second Feigenbaum constant (sequence [A006891] in OEIS),

α=2.502907875095892822289902873218.......

is the ratio between the width of a <u>tine</u> and the width of one of its two subtines (except the tine closest to the fold).
These numbers apply to a large class of <u>dynamical systems</u>. Both numbers are believed to be <u>transcendental</u> although have not been proven to be so.
http://encyclopedia.thefreedictionary.com/Feigenbaum+constant

Keith Briggs from the Mathematics Department of the University of Melbourne in Australia computed what he believes to be the world-record for the number of digits for the Feigenbaum number:

4.6692016091029906718532038204662016172581855774757686327456513430041343302113147371386897440239480138171659848551898151344086271420279325223124429888908908599449354632367134115324817142199474556443658237932020095610583305754586176522220703854106467494942849814533917262005687556659523398756038256372 25

4.66920160910299067185320382046620161725818557747576863274565134300413433021 1314 divided by 5184 = 0.0009006947548423978919469914777133876576501129586180109245265531140054271470315034<u>222222222</u>.........(continuously repeating digits out to infinity!).

Notice that when we multiply the repeating digits of 222222222, we get a value equal to 512, where 512 x 10.125 = 5184.

Miscellaneous Calculations

$$\frac{\pi^e}{5184}$$

Pi = 3.14159265358979323846264338327950288419716939937510582097494459230781 64 raised to the power of e = 2.71828182845904523536028747135266249775724709369995749669676277240766303535

Equals:

π^e = 22.4591577183610454734271522045437350275893151339966922492030025540669259182384944994456589468621582143209050370762 4567...

……….. Now when we divide π^e by 5184, we get the following:

22.4591577183610454734271522045437350275893151339966922492030025540669259182384944994456589468621582143209050370762 4567divided by 5184 =

0.0043323992512270535249666574468641464173590499872678804493061347519419224379318083525165237166014965691205449531397078 83873456790123456790......(continuously repeating digits out to infinity!).

Notice that when we multiply together the repeating digits of 123456790, we get the value equal to 45,360 = 8.75 x 5184.

$$\frac{\frac{1}{7}}{5184}$$

$\frac{1}{7} = 0.142857\ldots\ldots$(continuously repeating digits out to infinity!).

By dividing this result by 5184, we get:

0.00002755731922398589065255731922398589065255731922398589065 255731922398589065255731922398589065255731922398589065255731 905864197530……………..(continuously repeating digits out to infinity!). Notice too that the string of digits 731922398589065255 appears three (3) times before the final continuously repeating digits of 864197530 appear.

Notice that the product of the repeating digits of 864197530 = 181,440 = 35 x 5184 and the repeating digits of 731922398589065255= 17,010,000 x 5184.

$$\sqrt[2]{2^{\sqrt[2]{2}}}$$

$\sqrt{2} =$
1.4142135623730950488016887242096980785696718753 7694

Then
$2^{1.41421356237309504880168872420969807856967187537694}$ =
2.6651441426902. Then $\sqrt{2.6651441426902}$ = 1.6325269194382

Thus, $\sqrt[2]{2^{\sqrt[2]{2}}}$ = 1.6325269194382. When 1.6325269194382 is divided by 5184 we get
0.000314916458224961419753 08641975…..(continuously repeating digits out to infinity!). When the digits of 308641975 are multiplied, we get 181,440 = 35 x 5184.

$$\frac{\pi^\pi}{5184}$$

$\pi^\pi =$

36.46215960720791177099082602269212366636550840222881873 87093359229340738772908311728526015999781172131101719930 77612279............(variable digits out to infinity).

Then $\frac{\pi^\pi}{5184} =$

0.007033595603242266931132489587710672003542729244257102 38015226387402277659669962021081261604937849483277588193 1091073069058641975308........(continuously repeating digits out to infinity!).

Notice that the product of the repeating digits of 864197530 = 181,440 = 35 x 5184.

Pell's sequence (Last modified Tuesday, January 11, 2000)

Taken from:
http://www.cs.arizona.edu/patterns/sequences/pell.html

1,2,5,12,29,70,169,408,985,2378,5741,13860,33461,80872,195025,

470832,1136689,2744210,……..(this sequence is carried out for much further; see the website listed above).

If we take the series of numbers from the above example and then remove the "spaces" between the numbers and divide by 5184, we get a very interesting result:

1251229701694089852378574113860334618087219502547083211366892744210

Divided by 5184 =

2.41363754184816715350805191716885535894718268238249076266761
7176331018<u>518</u>……..(The digits <u>518</u> are continuously repeated out to infinity!).

Notice that the repeating digits of <u>518</u> are missing the number 4 from 5184.

What happens when we divide 1000 digits of Pi by 5184?

We will not list the 1000 digits of Pi in this book because these digits are readily available on the worldwide web.

The result of this division by 5184 =

0.000606017101............and continues with variable digits out to the 1008th decimal place)...and then is followed by the continuously repeating digits of 518518518 out to infinity!

What happens when we divide 70 digits of Pi divided by 5184?

The result of this division by 5184 =

0.000606017101386919991987392627947434969945441627965876894478191472281600385802469135802469......(where the digits 135802469 are continuously repeating out to infinity!).

Here we clearly see another pattern of repeating digits out to infinity? Note that the repeating digits 135802469 are the same repeating digits that we get when the "Total Atomic Weights of the 92 Natural Elements" are divided by 5184. This is truly amazing! So I think we can conclude that 5184 and 10368 and 2701 (the God Numbers) control the distribution of the digits of Pi and the pattern of repeating digits can vary depending on the number of Pi digits that are being divided into by these God Numbers.

Dr Nikolay Kosinov's research paper entitled: "Connection of Three Major Constants: Fine Structure Constant (α), the number Pi (π) and the Gold Ratio (φ). See unitron.com.ua.

Dr Kosinov thinks that he has found a mathematical connection with these three (3) constants with this equation:

$$\alpha^{20} = \sqrt[13]{\pi \phi^{14}} \times 10^{-43}$$

Thus, Dr Kosinov believes that the true value of alpha (α) is:

0.0072973519973773616957353015309841l.......

Meanwhile the CODATA 2006 value for alpha (α) is reported to be:

0.00729735253765..........

Thus, there is a difference of 0.000000000540273 between these two values.

Meanwhile, if you take the "total number of days in a Solar Year" to be equal to 365.2421896698 and divide it by 51840, you get a value for alpha (α) = 0.007045566930359. Thus, there is a difference of just 0.000251785067019 between my calculated value using the "total number of days in a Solar Year" and Dr Kosinov's mathematical results! It appears to me that there is a definite relationship between the "total number of days in a Solar Year" and alpha (α). Here again we see the amazing characteristics of the number 5184 and the other God numbers.

Conclusion

As I have demonstrated dozens of times throughout this book, that the numbers 5184 and 10368 and any of their permutations seem to be the "cement" of the universe. Obviously, with this small book it is impossible to analyze all aspects of our physical, mathematical and spiritual universe. I ask that scientists and mathematicians and theologians everywhere continue to study other aspects of our universe to confirm my findings.

I am not a professional scientist or mathematician or theologian, but it is evident to me that there is just ONE TRUTH in this universe and it encompasses the physical, mathematical and spiritual. There can be NO CONFLICT between these elements of the universe. As I have shown in this book, these physical, mathematical and spiritual aspects of the universe seem to all be related to the numbers 5184 and 10368. Moreover, these numbers coordinate directly with the Holy Bible scripture references to "God" where "His Name and Title" is the "Lord Jesus Christ". Remember: Jesus did say, "I am the WAY, the TRUTH and the LIFE". TRUTH is not just a complete and accurate set of facts, but is also the very essence of a person, the "Lord Jesus Christ".

Some readers may not want to agree with this last statement, but I think that the "math calculations" in this book leave us with no other choice. This has not been a book of "math theory", but a book of actual calculations. If my calculations are correct, then the truth should be obvious.

There is a very curious fact about the number 5184: I am told that "there are 5184 ways to place two (2) non-attacking Rooks on a chessboard (9 x 9)"! No wonder that chess is such a challenging game! God must truly be a master chess player.

As I stated earlier in this book, there are exactly 24 permutations of the number 5184. These 24 permutations add up to a value of 119988. Note that if we multiply the digits of 119988, we get a value of 5184. Furthermore, if we divide 119988000 by 37,875, we get the value 3168 = "Lord Jesus Christ" in Greek numerics! Thus, we see in this book that 5184 is a function of many different mathematical and physical measurements, as well as directly related to the value 3168.

We must remember that the Holy Bible was given to us in the Hebrew and the Greek languages. The ancient Hebrews and Greeks used their alphabet letters for words and numbers. Thus, when we look at the Holy Bible, we see thousands of "words" and thousands of "numbers". We could say that "God" communicated with us in "language" and in "numbers". We could call the Holy Bible a "digital signal from God". If God really did talk to us by "digital signal", then this "information" should be just as valid as the "language". In this computer age, we know that just "one wrong digit in a computer program" can create errors or complete shut down. If we are going to "find the truth" in science, mathematics or theology, we must be very precise in using the language and numbers. I think everyone can agree with that.

Bibliography

- E.W.Bullinger, Number in Scripture, Kregel Publications, Grand Rapids, MI USA 1967
- Michael Drosin, The Bible Code, Simon & Schuster, New York, NY USA 1997
- Bonnie Gaunt, Jesus Christ, The Number of His Name, Adventures Unlimited Press, Kempton, IL USA 1998
- Del Washburn & Jerry Lucas, Theomatics, Stein & Day, New York, NY USA 1977
- Bonnie Gaunt, Time and the Bible's Number Code, Adventures Unlimited Press, Kempton, IL USA 2001
- Bonnie Gaunt, Beginnings: The Sacred Design, Bonnie Gaunt, Jackson, MI USA 1995
- Bonnie Gaunt, The Bible's Awesome Number Code, Adventures Unlimited Press, Kempton, IL USA 2000
- Bonnie Gaunt, Nile: The Promise Written In Sand, Adventures Unlimited Press, Kempton, IL USA 2005
- Bonnie Gaunt, Apocalypse..And The Magnificent Sevens, Adventures Unlimited Press, Kempton, IL USA 2002
- Bonnie Gaunt, Genesis One: The Sacred Code, Adventures Unlimited Press, Kempton, IL USA 2003
- Bonnie Gaunt, The Stones and the Scarlet Thread, Adventures Unlimited Press, Kempton, IL USA 2001
- Bonnie Gaunt, Stonehenge and the Great Pyramid, Adventures Unlimited Press, Kempton, IL USA 1997
- Bonnie Gaunt, Jordan: The Fulfilled Promise, Adventures Unlimited Press, Kempton, IL USA 2006

- James Harrison, The Pattern & The Prophecy: God's Great Code, Isaiah Publications, Peterbourgh Ontario, Canada 1995

- Dr Peter Bluer, PhD, 373: A Proof Set In Stone, Lexis Hannah Publishing, Manchester, England 2001
- Jonathan Lawton Hamilton, The God-Code And Creation, Word of God, Inc, Anderson, SC USA 2004
- David Darling, The Universal Book of Mathematics, John Wiley & Sons, Inc, Hoboken, NJ USA 2004
- Glenn James and Robert C James, Mathematics Dictionary, D Van Nostrand Company, Princeton, NJ USA 1967
- Mark A Stevens, Editor, Merriam-Webster's Collegiate Encyclopedia, Merriam-Webster, Inc, Springfield, MA USA 2000
- Daniel W Matson, Signs of the End: A Discovery of Biblical Timelines, Inspiration Press, Fountain Valley, CA USA 2004
- IUPCA Commission on Atomic Weights and Isotropic Abundances, Atomic Weights of the Elements 2007, http://www.chem.qmul.ac.uk/iupac/AtWt/index.html
- Vernon Jenkins MSc, The Ultimate Assertion: Evidence of Supernatural Design In The Divine Prologue (Genesis 1:1), http://homepage.virgln.net/vernon.jcnkins/Evidences.htm , 2003
- John Tng, The God of Light of Genesis 1:1, http://www.fivedoves.com/, 2005. Plus many other articles too numerous to mention.

- Richard Amiel McGough, The Bible Wheel: A Revelation of the Divine Unity of the Holy Bible, Bible Wheel Ministries, Yakima, WA USA 2006. This is an excellent book!

- Dr Nikolay Kosinov, Connection of Three Major Constants: Fine Structure Constant (a), Pi (π), and Golden Ratio (ϕ), http://www.sciteclibrary.ru/eng/catalog/pages/2017.html, Publishing date: August 17, 2001
  Source: SciTecLibrary.ru
- Dr Chuck Missler, The Mysteries of Pi and e, http://www.khouse.org/articles/2003/482/, Koinonia House Online, 2007
- Dr Chuck Missler, Cosmic Codes: Hidden Messages from the Edge of Eternity, Koinonia House, Coeur d'Alene, Idaho USA 2004
- John J Parsons, Hebrew Consonants: Numeric Values of Hebrew Letters, http://www.hebrew4christians.com/index.html, Hebrew For Christians, 2007
- Tim Warner: Author, The Last Trumpet: Post-Trib Research Center, The Numeric Value of the Greek Alphabet, http://www.geocities.com/~lasttrumpet/greek.html, 2007
- Waldo N Larson, Gematria and Christmatics, http://bibleprobe.com/gematria.htm, 2001
- Dennis Lee Oberholtzer, Theophysic Gematria, published by Paradise Living, Paradise, PA, 2004
- Welcome to the World of Hummingbirds, see http://www.worldofhummingbirds.com/facts.php

- Nuclear Explosion Facts, see http://www.geocities.com/CapeCanaveral/Lab/2742/numbers.htm
- Koinonia House Online, see http://www.khouse.org/articles/2003/482.
- Pell's sequence, Last modified Tuesday, January 11, 2000 5:04 pm, http://www.cs.arizona.edu/patterns/sequences/pell.html
- Please note that I have used http://ttmath.slimaczek.pl/online_calculator for many of the examples of this book.

About the Author

Charles David Landis

Mr Landis studied Liberal Arts at Eastern Mennonite College and Bible and Theology at Eastern Mennonite Seminary. Later he graduated with a BA degree in Mathematics from Millersville University and pursued a short career in Information Technology. He later furthered his education with the American College where he earned the prestigious designation as a Chartered Financial Consultant. For the past thirty-eight years he as counseled numerous clients in all aspects of financial plans. He is currently President and CEO of Capital Assets Corporation, a financial services firm located in Southeastern Pennsylvania. In addition to his work in the financial industry, he has been an adult Bible teacher at several churches for the past 50 years.

Made in the USA
San Bernardino, CA
14 December 2012